Mathematics Problem Solving COACH

and
Applications

C

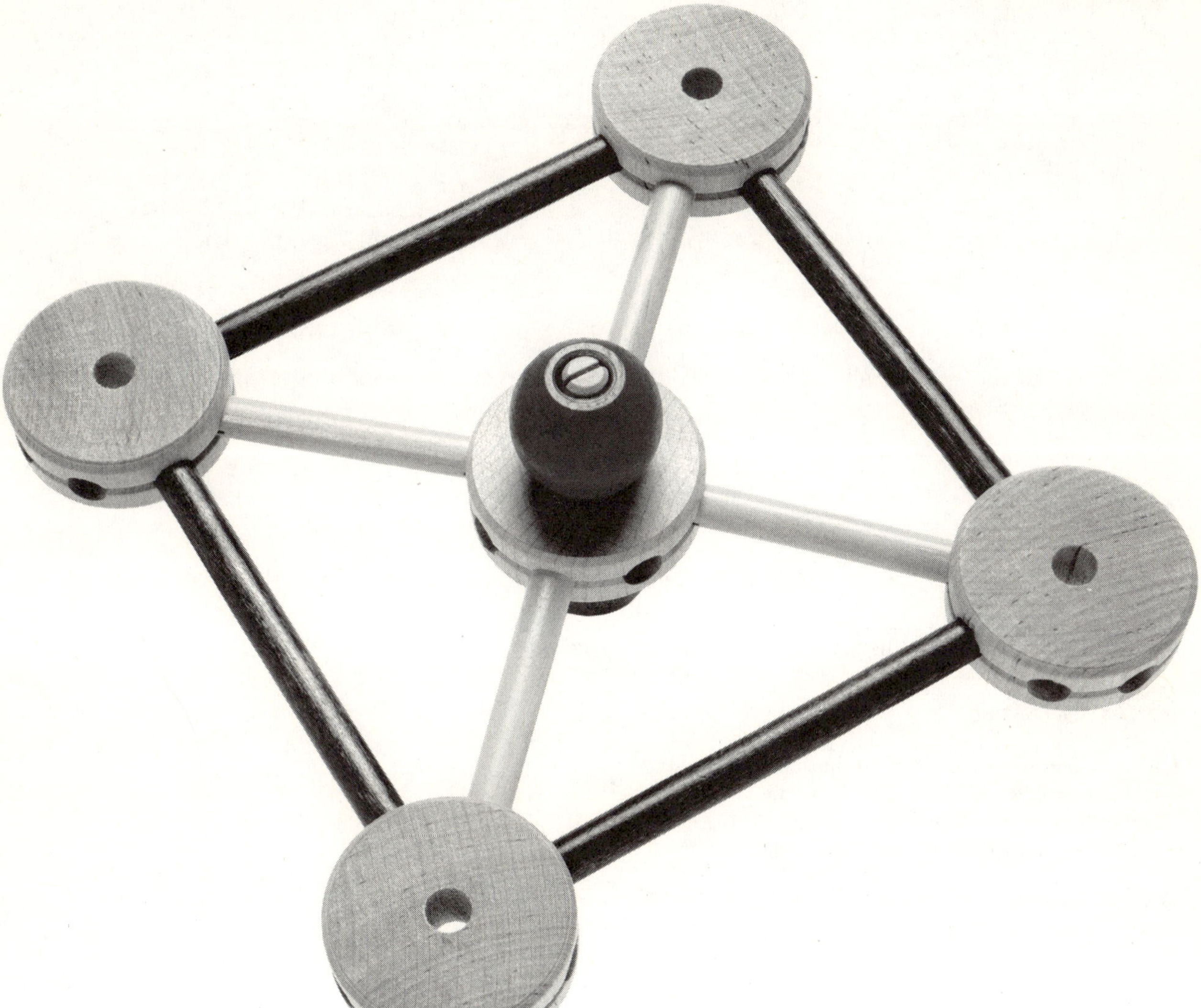

Mathematics Problem Solving Coach, Level C: Strategies and Applications

63NA
ISBN# **1-58620-539-0**

Editorial Contributors: Constance Lehmann, Dena Lindsay
Interior Design: Rocio Paez
Cover Design: Farzana Razak
Cover Photo: Tony Rinaldo/Index Stock Imagery

Triumph Learning 333 East 38th Street, New York, NY 10016-2777

A Haights Cross Communications company

Printed in the United States of America.

10 9 8 7 6 5 4 3 2 1

Table of Contents

To the Teacher 5
To the Student 6
Problem-Solving Strategies 7

I. Strategies for Solving Problems

Lesson 1 Draw a Picture 8
Lesson 2 Make a Model or Act Out 10
Lesson 3 Make an Organized List 12
Lesson 4 Make a Table 14
Lesson 5 Look for a Pattern 16
Lesson 6 Predict and Test 18
Lesson 7 Use Logical Thinking 20
Lesson 8 Work Backward 22
Lesson 9 Solve a Simpler Problem 24
Lesson 10 Write a Number Sentence 26

II. Applying the Strategies

Solving Problems with Whole Numbers

Lesson 11 Add and Subtract 28
Lesson 12 Multiply and Divide 32
Lesson 13 Money 36
Lesson 14 Number Theory 40

Solving Problems with Fractions and Decimals

Lesson 15 Fractions and Decimals 44

Lesson 16 Add and Subtract Like Fractions 48

Lesson 17 Add and Subtract Decimals 52

Solving Problems with Geometry and Measurement

Lesson 18 Metric and Customary Measurement 56

Lesson 19 Time and Temperature 60

Lesson 20 Plane Figures 64

Lesson 21 Solid Figures 68

Solving Problems with Algebra, Data, and Probability

Lesson 22 One-Step Equations 72

Lesson 23 Ordered Pairs on a Coordinate Grid .76

Lesson 24 Analyze Data 80

Lesson 25 Probability 84

III. Practice Test

Answer Sheet 88

Test-Taking Tips 89

Practice Test 90

To the Teacher

Mathematics Problem Solving Coach, Level C has been created to help your third-grade students improve their performance on your state proficiency tests.

The first part of the book focuses on ten important problem-solving strategies:

Strategy 1:	Draw a Picture
Strategy 2:	Make a Model or Act Out
Strategy 3:	Make an Organized List
Strategy 4:	Make a Table
Strategy 5:	Look for a Pattern
Strategy 6:	Predict and Test
Strategy 7:	Use Logical Thinking
Strategy 8:	Work Backward
Strategy 9:	Solve a Simpler Problem
Strategy 10:	Write a Number Sentence

The second part of the book focuses on using these problem-solving strategies to solve problems like those your students will have to solve on the state test. These problems cover the range of third-grade math topics, including whole numbers, money, fractions, decimals, geometry, measurement, algebra, data, and probability.

The third part of the book provides test-taking tips, a seven-page practice test, and an answer sheet. You may wish to have students use the answer sheet to practice how to record answers.

Most teachers use *Mathematics Problem Solving Coach, Level C* in one of two ways:

- Have students proceed straight through Lessons 1 to 25. Then give students the Practice Test.
- Begin by teaching the first ten lessons and then use the remaining lessons in conjunction with classroom study of that topic. Some students may need to use only those lessons in which they need further work. Then give students the Practice Test.

We hope and expect that by using this book, students will not only enhance their math competencies and do better on the state proficiency tests, but also that they will enjoy the problem-solving situations chosen.

To the Student

This book, ***Mathematics Problem Solving Coach, Level C,*** will help you learn how to solve word problems. It will prepare you for your state math test.

This book contains three parts:

Part I **Strategies for Solving Problems**

In this part of the book, you will learn 10 strategies to help you solve problems. Each strategy shows you how to use the four problem-solving steps:

- Find Out
- Plan
- Solve
- Look Back

Part II **Applying the Strategies**

In this part of the book, you will practice using the 10 strategies as you solve problems.

Part III **Practice Test**

In this part of the book, you will find tips for doing well on tests. You will take a test to practice your test-taking skills.

Mathematics Problem Solving Coach, Level C contains multiple-choice questions and open-ended questions. For the multiple-choice questions, there is only one correct answer. For the open-ended questions, you must write your own answer. For other questions, you will record your answer in a bubble grid.

Here is how you can use this book to prepare for your test:

1. In each lesson, read the Examples several times. This will help you see the steps you need to get the correct answer.
2. Try the sample test questions at the end of each lesson. If you cannot do some of them, study the Examples again.
3. As you work on Lessons 11–25, look back at Lessons 1–10 for problem-solving strategies you can use.

Good luck!

Ten Problem-Solving Strategies You Can Use To Be a Better Problem Solver

1 Draw a Picture
You can do this when you need to see the information in a problem to be able to solve it.

2 Make a Model or Act Out
You can do this when you need to see the data. Then you can watch how the solution is found.

3 Make an Organized List
You can do this to help you review the information in a problem. It will help you organize your thinking about it.

4 Make a Table
You can do this to organize information in a problem. Tables can help you see relationships and patterns in data.

5 Look for a Pattern
You can do this to predict what comes next. If you can find the rule for a pattern, you can use it to solve the problem.

6 Predict and Test
You can do this when it is difficult to work out the answer to a problem. You can make a guess and then test it. If your guess is not correct, use that guess to make a better guess.

7 Use Logical Thinking
You can do this when you need to think about how the information you know fits together.

8 Work Backward
You can do this when you know the total but need to find a missing part. Start at the end and work back to find the answer.

9 Solve a Simpler Problem
You can do this when the numbers in a problem are very big. Change them to smaller numbers that are easier to work with. Decide how to solve the problem. Then solve the original problem the same way.

10 Write a Number Sentence
You can do this when you need find a missing amount to solve a problem.

1 Draw a Picture

Sometimes drawing a picture can help you solve a problem.

Example

The art room has 2 square tables. There are 4 children at each table. Each child has 3 jars of paint. How many jars of paint are there in all?

Find Out

Think: *What do you know?*

- 2 tables
- 4 children at each table
- 3 jars of paint for each child

Think: *What do you need to find out?*

the number of jars of paint in all

Plan

You can *draw a picture.*

Then you can count all the jars of paint.

Solve

- Draw two squares.
- Draw 4 circles around each table for the children.
- Draw 3 Xs in front of each child.

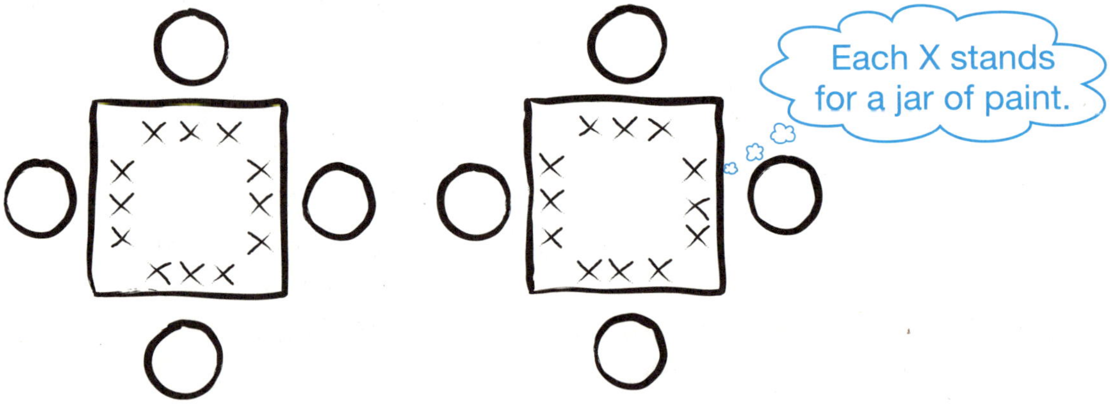

- Now count the Xs.

Answer: There are 24 jars of paint in all.

Look Back

Another way to solve this problem is by multiplying.

2 tables × 4 children = 8 children

8 children × 3 jars of paint = 24 jars of paint ✔

Problem Solving Practice

Find Out • Plan • Solve • Look Back

Draw a picture to solve each problem. Show your work.

1. Harry created a sign for the library club. He drew 3 books. Then he drew 5 stars on the front of each book. How many stars did he draw?

2. Josie painted a design on her T-shirt. The design had a triangle inside a square. The square was inside of a circle. List the shapes in order from the outside to the inside.

3. The monkey house is just inside the main entrance to the zoo. The giraffe fields are between the lion country and the polar bear pool. The lion country is between the giraffe fields and the monkey house. List the exhibits in order, starting at the entrance.

4. Kali, Miguel, Shanda, and Paul are sitting in a row. Shanda sits in the middle chair. Miguel sits 2 chairs to the left of Shanda. Paul sits 3 chairs to the right of Miguel. Kali sits between Miguel and Shanda. Who sits to the right of Shanda?

5. A bicycle has 2 wheels. A tricycle has 3 wheels. There are some bicycles and tricycles on the sidewalk. There are a total of 10 wheels. How many bicycles are there? How many tricycles are there?

2 Make a Model or Act Out

You can solve some problems by making a model or acting out the problem.

Example

Tina uses 3 toothpicks to make a triangle. She has 27 toothpicks. How many triangles can she make?

Find Out

Think: *What do you know?*

- 3 toothpicks make 1 triangle.
- Tina has 27 toothpicks.

Think: *What do you need to find out?*
the number of triangles Tina can make

Plan

You can *use* toothpicks to *act out* the problem.

Solve

Use 27 toothpicks. If you do not have toothpicks, use 27 craft sticks. Or use 27 crayons.

Make as many triangles as you can.

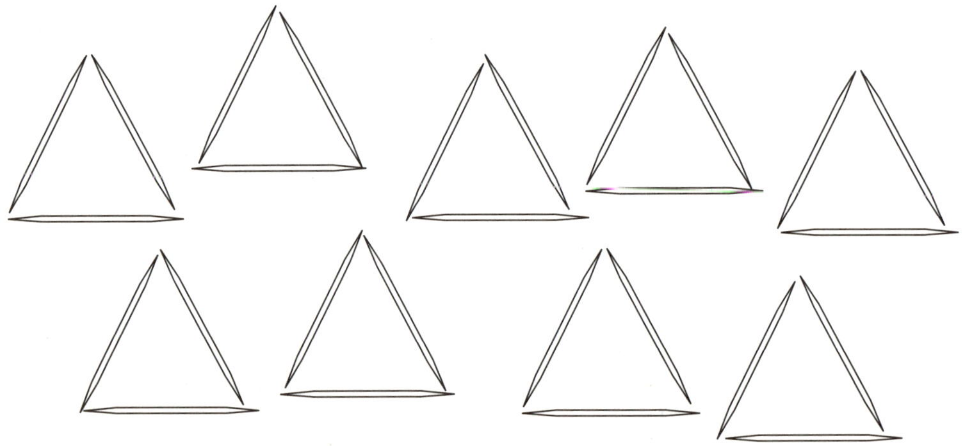

How many triangles can you make?

Answer: Tina can make 9 triangles.

Look Back

You can check your answer by adding.

3 + 3 + 3 + 3 + 3 + 3 + 3 + 3 + 3 = 27 toothpicks ✔

9 triangles

Problem Solving Practice

Find Out • Plan • Solve • Look Back

Make a model or act out to solve each problem. Show your work.

1. 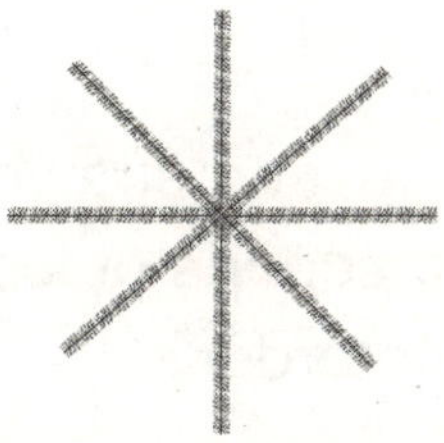

George can make a star using 4 pipe cleaners. How many pipe cleaners does he need to make 5 stars?

2. Darlene has 24 quarters. How many stacks of 6 quarters each can she make?

3. Frank and Liz are reading the same book. Frank has read $\frac{3}{4}$ of the book. Liz has read $\frac{2}{3}$ of the book. Who has read more of the book?

4. Yolanda has a jar of coins that she finds on the sidewalk. Last week, she found 5 quarters, 7 dimes, 3 nickels, and 14 pennies. What is the total amount she found?

5. Rachel has 32 stickers. She shares them with 3 friends. How many stickers does each person get?

6. Sherrie builds a tower of blocks for her little brother. The tower is 4 blocks long and 3 blocks wide. It is 6 layers high. How many blocks are in the tower?

3 Make an Organized List

Sometimes the best way to think about a problem is to make a list.

Example

An ice cream truck sells vanilla, chocolate, and strawberry ice cream. Mario wants a cone with two scoops of ice cream. How many different cones can he order?

Find Out

Think: *What do you know?*

- A truck sells vanilla, chocolate, and strawberry ice cream.
- Mario wants a cone with two scoops.

Think: *What do you need to find out?*

how many different two-scoop cones there are

Plan

You can *make an organized list* to find out.

Solve

List all the different ways to make a two-scoop cone.

vanilla-chocolate

vanilla-strawberry

vanilla-vanilla

chocolate-chocolate

chocolate-strawberry

~~chocolate-vanilla~~ — This is the same as vanilla-chocolate.

~~strawberry-chocolate~~ — This is the same as chocolate-strawberry.

strawberry-strawberry

~~strawberry-vanilla~~ — This is the same as vanilla-strawberry.

Answer: Mario can order 6 different cones.

Look Back

You can draw a picture to check your answer.

Count 6 different cones. ✔

Problem Solving Practice

Find Out • Plan • Solve • Look Back

Make a list to solve each problem. Show your work.

1. Ling, Tyrone, Carl, and Beth are choosing study partners. In how many different ways can they pair up?

2. You have these number cards.

3	**4**	**5**

How many different 3-digit numbers can you make?

3. The soccer team is getting a new uniform. The shorts can be tan, black, or white. The shirts can be yellow, red, or purple. How many different uniform choices does the team have?

4. How many ways can you make 25¢ using dimes, nickels, and pennies? Show all the ways.

Use the table for problems 5 and 6.

Design-Your-Own-Pasta Bar

Choose one from each category.

Pasta	Sauce	Meat
spaghetti	red	meatballs
shells	white	sausage

5. Mario wants pasta with sauce. How many choices does he have?

6. Amy wants sauce and meat on her pasta. How many choices does she have?

4 Make a Table

Another way to think about a problem is to make a table of all the facts you know.

Example

The children's zoo has 35 parrots, 47 ducks, and 29 geese in the birdhouse. There are 22 snakes, 53 turtles, and 17 lizards in the reptile house. How many more birds does the zoo have than reptiles?

Find Out

Think: *What do you know?*
The number of parrots, ducks, geese, snakes, turtles, and lizards that are in the zoo

Think: *What do you need to find out?*
How many more birds there are than reptiles

Plan

You can *make a table* to sort the facts.

Solve

Sort the animals into birds and reptiles.

Birds	Reptiles
35 parrots	22 snakes
47 ducks	53 turtles
29 geese	17 lizards

Add the number of birds. 35 + 47 + 29 = 111

Add the number of reptiles. 22 + 53 + 17 = 92

Subtract. 111 − 92 = 19

Answer: The children's zoo has 19 more birds than reptiles.

Look Back

You can *make a model* to check your answer.
Use place-value blocks to show how many birds.
Then take away 22, 53, and 17 blocks for the reptiles.
You will have 19 left. ✔

Problem Solving Practice

Find Out • Plan • Solve • Look Back

Make a table to solve each problem. Show your work.

Use the following information for problems 1–3.

There are 3 third-grade classes. Mrs. Boston's class has 13 boys and 9 girls. Mr. Wright's class has 12 boys and 10 girls. Miss Hiroi's class has 14 boys and 8 girls.

1. How many boys are in third grade?

2. How many girls are in third grade?

3. How many more boys than girls are in third grade?

4. Brad can paint a room in 3 days. How many rooms can he paint in 15 days?

5. This list shows the favorite pet named by each student in the class. What is the favorite pet of most of the students?

dog	cat	rabbit	hamster
cat	fish	fish	cat
fish	cat	cat	dog
rabbit	dog	dog	cat

6. Student tickets to the circus cost $6.50 each. How much will 6 tickets cost?

5 Look for a Pattern

You can solve some problems by looking for a pattern.
If you can find a rule, you can use it to solve the problem.

Example

Maxine is making this chain of shapes.

If she follows the pattern, what 5 shapes will she use next?

Find Out

Think: *What do you know?*
The chain has hearts and diamonds.

Think: *What do you need to find out?*
the next 5 shapes in the chain

Plan

Look for a pattern in the shapes Maxine used.

Solve

Describe the pattern in words.

heart-heart-diamond-diamond-diamond-heart-heart-diamond-diamond-diamond

2 hearts, 3 diamonds, 2 hearts, 3 diamonds, 2 hearts, 3 diamonds

Use the rule to extend the pattern.

Answer: Maxine will use 2 hearts and 3 diamonds next.

Look Back

Draw the chain. Extend the pattern.

Problem Solving Practice

Find Out • Plan • Solve • Look Back

Look for a pattern to solve each problem. Show your work.

1. Tom is painting this design on his kitchen wall. If he continues this pattern, what figure should he paint next?

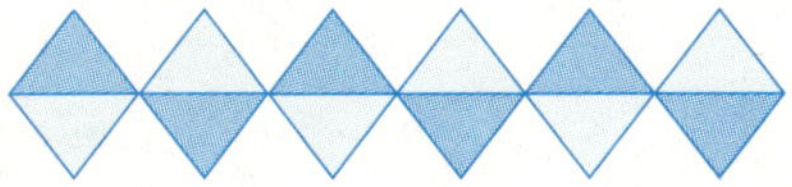

2. There are 4 quarters in \$1.00 and 8 quarters in \$2.00. How many quarters do you need to have \$4.00?

3. The teams are ready to run the 3-legged race. Each team has two people. Joe counts everyone who is in the race. He skip counts by 2. He counts to 16. How many teams does he count?

4. Mrs. Ross pasted a line of shapes above the chalkboard. One of the shapes is hidden by the flag. Which shape is hidden?

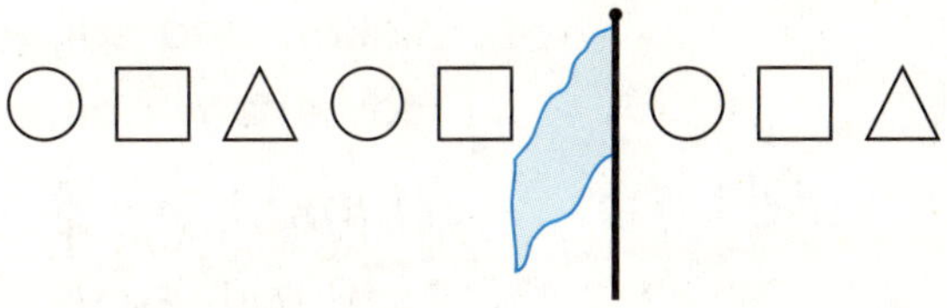

5.

After-School Bus Schedule	
Bus 1	2:30 P.M.
Bus 2	2:45 P.M.
Bus 3	3:00 P.M.
Bus 4	3:15 P.M.
Bus 5	3:30 P.M.
Bus 6	?

If the pattern continues, what time does Bus 6 leave the school?

6 Predict and Test

Sometimes you can guess the answer and then check to see if you are right. If your guess is wrong, you can guess again. This is called predict and test.

Example

Tony has 70¢ in his pocket. He has 9 coins. Some are nickels and some are dimes. How many coins are dimes? How many coins are nickels?

Find Out

Think: *What do you know?*

- Tony has 70¢ in nickels and dimes.
- He has 9 coins.

Think: *What do you need to find out?*
how many of each coin Tony has

Plan

You can *predict and test* to solve this problem.
Each guess must use 9 coins.

Solve

Try 6 dimes and 3 nickels.
$6 \times 10¢ \qquad 3 \times 5¢$
$60¢ \quad + \quad 15¢ = 75¢ \leftarrow$ Too high.
Guess again.

Try 4 dimes and 5 nickels.
$4 \times 10¢ \qquad 5 \times 5¢$
$40¢ \quad + \quad 25¢ = 65¢ \leftarrow$ Too low.
Guess again.

Try 5 dimes and 4 nickels.
$5 \times 10¢ \qquad 4 \times 5¢$
$50¢ \quad + \quad 20¢ = 70¢ \leftarrow$ That's it!

Answer: Tony has 5 dimes and 4 nickels.

Look Back

You can also use play money to check your answer.

9 coins = 70¢

Problem Solving Practice

Find Out • Plan • Solve • Look Back

Predict and test to solve each problem. Show your work.

1. Jamie bought a comic book for 53¢. She gave the clerk 17 coins. All the coins were dimes and pennies. How many of each coin did she use?

Use this price list for problems 2 and 3.

Apples	25¢ each
Pears	30¢ each
Peaches	35¢ each

2. Pam bought 2 pieces of fruit at the fruit stand. She spent 65¢. What did she buy?

3. Chris bought some apples and some pears. He spent $1.35. How many of each did he buy?

4. Mrs. Patton has twice as many chickens as cows on her farm. She has a total of 24 chickens and cows. How many of each does she have?

5. Halley is older than Sandy. The sum of their ages is 38, and the difference in their ages is 4. How old is Halley?

6. Ed has a length of rope that is 23 feet long. He cuts it into 2 parts. One part is 5 feet longer than the other part. How long are the 2 parts of rope?

7 Use Logical Thinking

Sometimes you can use logical thinking to solve a problem.

Example

Maria's baseball jersey has a 2-digit number on it. The number is less than 60. Both of the digits are odd numbers. The sum of the digits is 14. What is Maria's number?

Find Out

Think: *What clues do you have?*

- 2-digit number less than 60
- Both digits are odd.
- Sum of the digits is 14.

Think: *What do you need to find?*
What number fits the clues?

Plan

You can *use logical thinking* to find the number.
Use the clues.

Solve

Make a list of all the 2-digit numbers with odd digits.
Stop before 60.

1 + 1 = 2	3 + 1 = 4	5 + 1 = 6
1 + 3 = 4	3 + 3 = 6	5 + 3 = 8
1 + 5 = 6	3 + 5 = 8	5 + 5 = 10
1 + 7 = 8	3 + 7 = 10	5 + 7 = 12
1 + 9 = 10	3 + 9 = 12	5 + 9 = 14

Which digits have a sum of 14?
Answer: Maria's number is 59.

Look Back

Is your answer reasonable?
Is 59 a 2-digit number? ✔
Are 5 and 9 odd digits? ✔
5 + 9 = 14 ✔

Problem Solving Practice

Find Out • Plan • Solve • Look Back

Use logical thinking to solve each problem. Show your work.

1. Gary is older than 10 and younger than 30. His age is an even number. The ones digit of his age is equal to 3 times the tens digit. How old is Gary?

2. Janelle makes yarn dolls to sell. She is packing them in boxes to send to a craft fair. She can pack 4 dolls in each box. How many boxes will she need to pack 14 dolls?

3. Jen is 4 years older than Sam. Sam is 3 years younger than Randy. Randy is 10 years old. How old is Jen?

4. Thor is shorter than Rover. Rover is taller than Fido. Hugo is shorter than Thor. Which dog is tallest?

5. Fred, Larry, and Bob went on a fishing trip. Larry caught 6 more fish than Fred. Fred caught twice as many fish as Bob. Larry caught 8 fish. How many fish did Bob catch?

6. Aden, Tina, and Cal get together to practice their musical instruments. Aden sits in the first chair. Cal sits at the end. He does not play the violin. The one who sits in the middle plays the flute. Who plays the violin?

8 Work Backward

What if you need to find a missing part to solve the problem?
You can start at the end and work backward.

Example

Jamal spends \$20.00 at the bookstore. He buys a poster for \$4.50. He buys magazines for \$7.00. He buys his sister a picture book. How much does the picture book cost?

Find Out

Think: *What do you know?*

- Jamal spends \$20.00.
- 1 poster costs \$4.50.
- The magazines cost \$7.00.

Think: *What do you need to find out?*

how much Jamal spends for a picture book

Plan

You can *work backward* to find the answer.

Solve

Start by finding out how much the poster and the magazines cost.

\$4.50 + \$7.00 = \$11.50

Then subtract that from the amount Jamal spends.

\$20.00 − \$11.50 = \$8.50

Answer: The picture book costs \$8.50

Look Back

Does the answer make sense?
Check by working forward.

poster + magazines + picture book = \$20.00

\$4.50 + \$7.00 + \$8.50 = \$20.00 ✔

Problem Solving Practice

Find Out • Plan • Solve • Look Back

Work backward to solve each problem. Show your work.

1. Nick bought a jar of paint for $6, a brush for $4, and a package of paper for his art project. He gave the clerk $20 and received $2 in change. How much did the paper cost?

2. Connie feeds her cat 1 cup of food every day. At the end of the week, there are 12 cups of cat food left in the bag. How much cat food was in the bag at the beginning of the week?

3. Rashid spent 2 hours washing his car, 3 hours mowing the lawn and 2 hours cooking dinner. He finished everything at 6 P.M. What time did he start?

4.

Train	Departs
101	7:00 A.M.
103	8:30 A.M.
105	10:00 A.M.
107	11:30 A.M.

Raul has a ticket for train 103. He wants to get to the train station an hour before the train leaves. It will take him 1 hour to drive to the train station. What time should he leave his house?

5. The temperature increased 12° between 9 A.M. and noon. The temperature decreased 5° between noon and 6 P.M. The temperature was 65°F at 6 P.M. What was the temperature at 9 A.M.?

9 Solve a Simpler Problem

Sometimes the numbers in a problem are very big. You can change them to smaller numbers and solve the simpler problem. Then you can solve the real problem the same way.

Example

The third, fourth, and fifth grades are going to see the play *Pinocchio*. There are 315 students in all. There are 9 buses. The same number of children will ride in each bus. How many children will ride in each bus?

Find Out

Think: *What do you know?*

- 315 children
- 9 buses

Think: *What do you need to find out?*
how many children will ride in each bus

Plan

You can *solve a simpler problem* first so you will know how to solve the real problem.

Solve

Try 15 children instead of 315.
Try 3 buses instead of 9.

$$\begin{array}{r} 5 \\ 3\overline{)15} \end{array}$$

Divide to find how many children on a bus.

Now solve the real problem the same way.

$$\begin{array}{r} 35 \\ 9\overline{)315} \end{array}$$

Answer: 35 students will ride on each bus.

Look Back

To check your answer, you can make a model.

You can use place value blocks to make 9 groups of 315. ✔

Problem Solving Practice

Find Out • Plan • Solve • Look Back

Solve each problem by solving a simpler problem first. Show your work.

1. The community center is taking 96 children to a basketball game. Each van can hold 8 children. How many vans will they need?

2. Selina has 10 pounds of nuts. She wants to put 4 ounces of nuts in each paper cup. How many paper cups can she fill?

1 pound = 16 ounces

3. Brian's dog is 1 yard 4 inches tall. Pamela's dog Rosie is 42 inches tall. Which dog is taller?

1 yard = 36 inches

4.

Shanique's tank holds 8 quarts of water. Lori's tank holds 15 pints of water. Which tank holds more water?

1 quart = 2 pints

5. Liza jogged for 1 hour 15 minutes. Noah jogged for 70 minutes. Who jogged for a longer time?

1 hour = 60 minutes

10 Write a Number Sentence

Sometimes you can write a number sentence to solve a problem.

Example

Many important things were invented before 2000.

Year	Invention
1885	Bicycle
1891	Zipper
1895	X-ray
1980	Rollerblades
1987	Laptop computer

How many years before Rollerblades was the bicycle invented?

Find Out

Think: *What do you know?*

- Rollerblades were invented in 1980.
- The bicycle was invented in 1885.

Think: *What do you need to find out?*
the number of years between those inventions

Plan

Write a number sentence to find the answer.

Solve

Think: *What operation will you use?*
Subtract to find how many years between two numbers. Write a subtraction sentence.

Use □ to stand for what you need to find out.
1980 − 1885 = □
□ = 95

Answer: The bicycle was invented 95 years before Rollerblades.

Look Back

You can write an addition sentence to check your answer.
1885 + 95 = 1980 ✔

Problem Solving Practice

Find Out • Plan • Solve • Look Back

Write a number sentence to solve each problem. Show your work.

1. The zipper was invented in 1891. Velcro was invented in 1948. How many years before Velcro was the zipper invented?

2. The airplane was invented in 1903. The helicopter was invented 36 years later. When was the helicopter invented?

3. Jeannie has 5 boxes of cookies. There are 6 cookies in each box. How many cookies does Jeannie have?

4. A horse has 18 more bones than a human. A human has 206 bones. How many bones does a horse have?

5. Manny has 4 pairs of blue socks and 3 pairs of green socks. How many socks does Manny have in all?

6. Mr. Chin's class made a table showing the number of their pets.

Pet	Number
Dog	18
Cat	25
Bird	7
Fish	12

How many pets do Mr. Chin's students have?

11 Add and Subtract

You can combine numbers by adding. You can compare two amounts or find their difference by subtracting.

Example 1

A cape cobra can grow up to 84 inches. How much longer can a king cobra grow?

Snake	Length (in inches)
Black Mamba	168
Asian Pit Viper	60
King Cobra	192

Find Out

You want to find out how much longer a king cobra is than a cape cobra. The table gives too much information. You need to know only the length of a king cobra. A king cobra can be 192 inches long.

Plan

You can *write a number sentence* to solve this problem.

Think: *First chose the operation you will use.*

You need to find how much more 192 is than 84. Use subtraction.

Solve

192 − 84 = □ *Write a number sentence.*

$$\begin{array}{r} 1\overset{8}{\not{9}}{}^{1}2 \\ -\ 84 \\ \hline 108 \end{array}$$

Regroup and subtract.

Answer: A king cobra can grow 108 inches longer than a cape cobra.

Look Back

You can use addition to be sure you subtracted correctly.

108 + 84 = 192

This is the length of a king cobra, so your answer is correct. ✔

Try It

Example 2

A Saint Bernard can weigh 198 pounds. A Clydesdale can weigh 2,152 pounds more than a Saint Bernard. How much can a Clydesdale weigh?

Find Out

You know:
- the weight of a Saint Bernard
- how much more a Clydesdale can weigh

You need to find out how much a Clydesdale can weigh.

Plan

You can *solve a simpler problem* first.

Solve

Rewrite the problem using easy numbers.

A Saint Bernard can weigh _____ pounds.

A Clydesdale can weigh _____ pounds more.

Solve the easier problem.

____

Think: Will you add or subtract? ☐ __________

Write the answer here __________

Follow the same steps to solve the original problem.

____

☐ __________

____

Answer: A Clydesdale can weigh _______ pounds.

Look Back

How can you use estimation to check your answer?

____

Practice Test Questions

Find Out • Plan • Solve • Look Back

1. In 2002, the National Baseball League's record for Runs Batted In was 128. That year, the American League record was 142 runs batted in. How much greater was the National League record? (Hint: *Write a number sentence.*)

A 270 more
B 260 more
C 24 more
D 14 more

2. Mt. St. Helens is an active volcano in Washington. It is 8,363 feet high. Mt. Pavlof is an active volcano in Alaska. It is 99 feet lower than Mt. St. Helens. How tall is Mt. Pavlof? (Hint: *Use logical thinking.*)

A 8,462 feet
B 8,364 feet
C 8,264 feet
D Not here

3. Sally is 3 years older than Rob. Kent is 4 years younger than Sally. Rob is 18 years old. How old is Kent? (Hint: *Use logical thinking.*)

A 17 years old
B 16 years old
C 15 years old
D 14 years old

4. Sandra has $122 in her savings account. She hopes to add $8 every month to her account. How many months will it take her to save a total of $170?

A 4 months
B 5 months
C 6 months
D 8 months

5. Mrs. Garcia pays for a $25 game controller and a $59 video game with a $100 bill. How many dollars will she get back?

Record your answer. Then fill in the bubbles. Be sure to use the correct place value.

⓪	⓪
①	①
②	②
③	③
④	④
⑤	⑤
⑥	⑥
⑦	⑦
⑧	⑧
⑨	⑨

6. Students in two of the classes listed below will go on a class trip to the Computer Show. The total number of students going on the trip will be 55. Which two classes will go on the trip? Explain your thinking. (Hint: *Predict and test.*)

Class	Number of Students
Mr. Wong	28
Ms. Carter	26
Mrs. Lamont	27
Ms. Moore	25

____

____

____

____

7. Ricky went to the mall. He shopped for 2 hours. Then he took 1 hour for lunch. After lunch, he went to a movie that lasted 2 hours. It took Ricky an hour to get home from the movie. If he got home at 4:30 P.M., at what time did he arrive at the mall? (Hint: *Work Backward.*)

8. Tyrone scored 4,088 points in a video game. The top score is 6,000 points. How many points away from the top score is Tyrone?

9. Tod has baseball and basketball trading cards. He has 10 more baseball cards than basketball cards. He has 38 cards in all. How many basketball cards does he have?

10. Mrs. Chao started knitting a scarf today. She knitted 24 rows. If she knits 1 more row each day than the day before, how many rows will she have knit in 4 days?

____

____

____

____

Strategies: Picture • Model/Act Out • Organized List • Table/Graph • Pattern • Predict & Test • Logical Thinking • Work Backward • Simpler Problem • Number Sentence

12 Multiply and Divide

When you multiply, you combine groups that are the same size. When you divide, you find how many are in each group or how many groups there are.

Example 1

The pictograph shows the number of books the students in each grade plan to read in a week. The third grade's goal is 100 books. How many weeks will it take the third grade reach its goal?

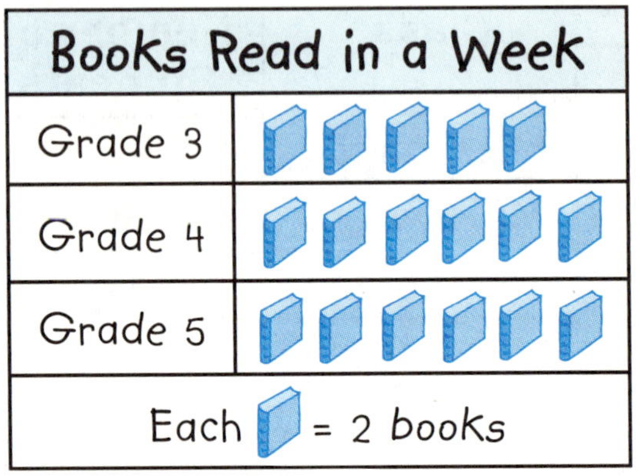

Find Out You want to find out how many weeks it will take the third grade to read 100 books. The table gives too much information. The third grade reads 10 books in 1 week.

Plan You can *write a number sentence* to solve this problem.

Think: *What operation will you use to solve the problem?* Since you know that the third grade reads 10 books in 1 week, you can divide 100 by 10 to find how many weeks it will take to read 100 books.

Solve $100 \div 10 = \square$ *Write a number sentence.*

$100 \div 10 = 10$ *Solve.*

Answer: The third grade will reach its goal in 10 weeks.

Look Back You can also *make a table* to solve this problem. Make sure to check your answer.

Week	1	2	3	4	5	6	7	8	9	10
Number of Books	10	20	30	40	50	60	70	80	90	100

✔

Try It

Example 2

The students in the drama club are setting up 16 seats. The club president wants the seats set up in equal rows. She does not want more than 10 rows. In how many different ways can the seats be set up?

Find Out

You know:
- the number of seats needed
- Each row will have the same number of seats.
- There will be no more than 10 rows.

You need to find the ways the 16 seats can be set up.

Plan

You can *make a model or act out* to solve the problem.

Solve

Use 16 cubes to model the seats. Set all the cubes in equal rows in as many different ways as you can. Here are two ways to do it.

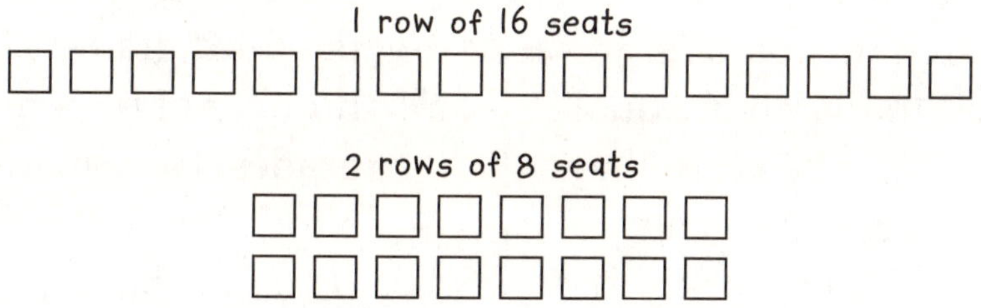

What other ways can you set up the seats in equal rows? Draw them here.

Answer: Seats can be set up __________ different ways.
1 row of 16; 2 rows of 8;

__________ rows of __________;

__________ rows of __________.

Look Back

How else could you have solved this problem?

____

Practice Test Questions

Find Out • Plan • Solve • Look Back

1. In 2002, four of the top Major League Soccer players each scored the same number of goals. The total number of goals scored by all 4 players was 116. How many goals did each player score? (Hint: *Write a number sentence.*)

A 464 goals
B 120 goals
C 112 goals
D 29 goals

2. The 64 students in the school band will march in a parade. Students need to march in equal rows. There cannot be more than 10 students in a row. Which arrangement will **not** work? (Hint: *Make a model or act out.*)

A 32 equal rows of students
B 16 equal rows of students
C 8 equal rows of students
D 4 equal rows of students

3. Jared earns $3 each day walking a neighbor's dog. He wants to save $40 to buy a computer game. What is the least number of days he must walk the dog to earn $40?

A 12 days
B 13 days
C 14 days
D 15 days

4. There are 24 students in the second grade and 28 students in the third grade. There is an equal number of boys and girls in both grades. How many boys are there altogether in the two grades?

Record your answer. Then fill in the bubbles. Be sure to use the correct place value.

⓪	⓪
①	①
②	②
③	③
④	④
⑤	⑤
⑥	⑥
⑦	⑦
⑧	⑧
⑨	⑨

5. The time it takes a planet to revolve around the sun is a year on that planet. The table shows how many Earth days are equal to 1 year on 3 different planets.

Approximate Length of a Year on Planets

Planet	Aproximate Time to Revolve Around the Sun
Earth	365 days
Venus	225 days
Mars	687 days

How long will it take the planet with the longest year to revolve 3 times around the sun? (Hint: *Solve a simpler problem.*)

6. There are 27 children who want to play baseball and 36 children who want to watch the game. Each team will have 9 players. What information is **not** needed to find how many baseball teams the children who want to play can make? (Hint: *Use logical thinking.*)

7 A large carton of oranges is 3 times as heavy as a medium carton. A medium carton is 2 times as heavy as a small carton. A small carton of oranges weighs 6 pounds. How much does a large carton of oranges weigh? Justify your answer.

8. Mark read a book in 4 days. He read the last 56 pages on the fourth day. If Mark read twice as many pages each day as he did the day before, how many pages did he read on the first day?

9. A toy store just got 24 model cars. The cars were delivered in 6 separate boxes. Each box has 2 bags of cars. Each bag has the same number of cars. How many cars are in each bag?

Strategies: Picture • Model/Act Out • Organized List • Table/Graph • Pattern • Predict & Test • Logical Thinking • Work Backward • Simpler Problem • Number Sentence

13 Money

Count, add, subtract, multiply, and divide money the same way as for whole numbers. Remember:

1 penny = 1¢ | 1 nickel = 5¢ | 1 dime = 10¢
1 quarter = 25¢ | 1 dollar = 100¢

Example 1

Amber has a combination of the following coins:

Amber has 55¢. What possible combinations of coins can she have?

Find Out You know Amber has at least 1 nickel and 1 dime. You want to find out the number of nickels and dimes she might have.

Plan You can *make an organized list* to solve this problem.

Think: *As you make your list, follow a pattern starting with 1 dime. Remember, all of your combinations must have a value of exactly 55¢.*

Solve

Increase the number of dimes by 1. Then decide how many nickels you need to make 55¢

1 dime, 9 nickels
2 dimes, 7 nickels
3 dimes, 5 nickels
4 dimes, 3 nickels
5 dimes, 1 nickel

Answer: Amber can have 5 possible coin combinations.

Look Back You can look for a pattern to be sure that your answer makes sense.

Check your list. It should show that each time the number of dimes increases by 1, the number of nickels decreases by 2. ✔

Try It

Example 2

Alex used the money in his savings account to buy a CD for $15 and a DVD for $22. Now he has $28 left in his account. How much did Alex have in his savings account to begin with?

Find Out

You know:
- the cost of the CD
- the cost of the DVD
- how much is left in the account

You need to find the amount Alex had in his account before he spent any money.

Plan

You can *work backward* to solve the problem.

Solve

This is a multistep problem. First show the order of the steps as they are given in the problem.

?	–	______	–	______	=	______
Amount in savings account		Cost of CD		Cost of DVD		Amount left

Now work backward.
Add the cost of the CD and the cost of the DVD to the amount left in the savings account. the sum is the amount Alex had to begin with.

______	+	______	+	______	=	______
Amount left in account		Cost of CD		Cost of DVD		Amount at the beginning

Answer: Alex started with __________ in his account.

Look Back

How can you check that your answer makes sense?

____

Practice Test Questions

Find Out • Plan • Solve • Look Back

1. Raul has a nickel, a dime, a quarter, and a penny. He will give his little brother 2 of the coins. What is the number of possible pairs of coins he can give to his brother? (Hint: *Make an organized list.*)

A 16 **C** 6

B 12 **D** 4

2. Tom bought a sandwich for $3.50 and a fruit salad for $1.25. He paid for lunch and got $5.25 change. How much did he give the clerk? (Hint: *Work backward.*)

A $4.75 **C** $8.75

B $6.50 **D** $10.00

3. Wendy divided her money equally among 4 charity boxes. She put 2 quarters, 1 dime, and 3 nickels in each box. How much money altogether did she put in the boxes? (Hint: *Make a model or act out.*)

A $4.25

B $3.00

C $2.75

D $2.00

4. Beads are on sale for 3 bags for $2.50. Meg needs 12 bags of beads to make bracelets for presents. How much will the beads cost?

A $10.00 **C** $15.00

B $12.50 **D** Not here

5. These items are for sale at a fair.

Ray buys the ball and glove. He pays $2.00. How much change should he get?

Record your answer. Then fill in the bubbles. Be sure to use the correct place value.

⓪	⓪
①	①
②	②
③	③
④	④
⑤	⑤
⑥	⑥
⑦	⑦
⑧	⑧
⑨	⑨

6. Teresa can buy 2 notebooks for $1.50, 3 notebooks for $2.25, and 4 notebooks for $3.00. How much would 6 notebooks cost? (Hint: *Look for a pattern.*)

7. An apple costs 35¢ cents in a vending machine. The machine will take dimes and nickels. What are the different coin combinations Erika can use to buy the apple? (Hint: *Make an organized list.*)

8. Pablo spent $5.50 on some slices of pizza and a bottle of juice for lunch. A slice of pizza costs $2.00. A bottle of juice costs $0.50 less than a slice of pizza. How many pizza slices did Pablo buy? (Hint: *Make a model or act out.*)

9. Grace earns $5.25 per week for walking a neighbor's child to school. Mitch earns $2.50 each time he washes a neighbor's car. Mitch washes two cars this week. Who earns more this week? How much more? Explain how you know.

10. Mike and his brother want to buy a hat for their mom.

Winter SALE!	
Scarf	$4
Gloves	$8
Knit Hat	$6

Mike has $2.85. His brother has $4.10. How can you estimate to tell if they have enough money?

Strategies: Picture • Model/Act Out • Organized List • Table/Graph • Pattern • Predict & Test • Logical Thinking • Work Backward • Simpler Problem • Number Sentence

14 Number Theory

You can use problem-solving skills to help solve problems that involve comparing and ordering, skip-counting, and factors and multiples.

Example 1

Students at the Acorn Elementary School put on a play. The table shows how many people came to see the play each night.

List the nights in order from greatest to least number of people.

School Play Attendance

Night	Number of People
Wednesday	198
Thursday	195
Friday	203
Saturday	200

Find Out You know how many people attended each night. You need to order the numbers from greatest to least.

Plan You can *draw a diagram or picture* to solve this problem.

Solve Draw a number line. Write the numbers from the table on the number line.

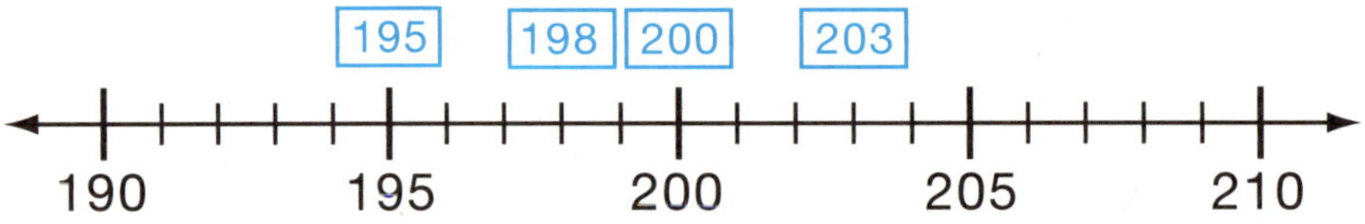

Write the numbers in order from greatest to least.

203 > 200 > 198 > 195

Friday, Saturday, Wednesday, Thursday

Answer: The nights in order from greatest to least attendance are Friday, Saturday, Wednesday, Thursday.

Look Back You can make a place-value table to compare and order the numbers. ✔

Try It

Example 2

Larinda bathes her dog George every 3 weeks. She changes the water in her fish tank every 2 weeks. Today, she bathed George and changed the fish tank water. In how many weeks will she have to do both again on the same day?

Find Out

You know:
- George is bathed every _____ weeks.
- Larinda changes the fish tank water every _____ weeks.
- Today Larinda did both jobs.

You need to find out when Larinda will do both jobs again on the same day.

Plan

You can *make an organized list* to solve the problem. Use multiples of 2 and multiples of 3 to find the least common multiple.

Solve

This is a multistep problem. Follow the steps in the correct order.

Multiples of 2:

2, _____, _____, _____ *List the multiples of 2.*

Multiples of 3:

3, _____, _____, _____ *List the multiples of 3.*

Circle common multiples.

The least common multiple is __________.

Answer: Larinda will bathe the dog and change the water on the same day in __________ weeks.

Look Back

How can you solve this problem another way?

____

Practice Test Questions

Find Out • Plan • Solve • Look Back

1. Beth is 45 inches tall. Mike is 51 inches tall, and Karin is 43 inches tall. Brent is 48 inches tall. What is their order from shortest to tallest? (Hint: *Draw a picture.*)

A Karin, Beth, Brent, Mike
B Mike, Brent, Beth, Karin
C Beth, Mike, Karin, Brent
D Not here

2. Every 3 days, the town picks up garbage. Garbage is picked up on February 3. What is the last day in February that garbage will be picked up? (Hint: *Make an organized list.*)

A February 6 **C** February 28
B February 27 **D** February 29

3. Which of the following is **not** true? (Hint: *Write a number sentence.*)

1 foot = 12 inches

A 3 feet > 24 inches
B 4 feet $= 48$ inches
C 5 feet > 50 inches
D 6 feet < 66 inches

4.

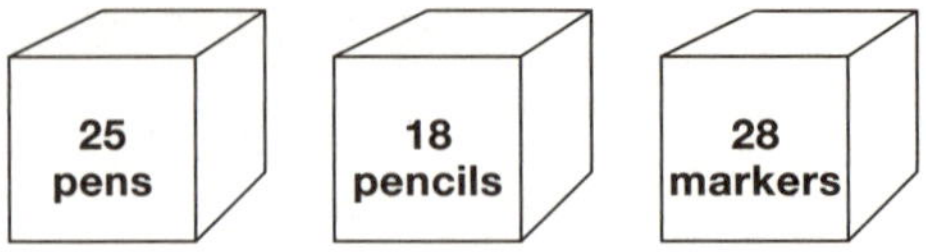

Which of the following shows the relationship among the number of items in each box?

A $25 > 18 > 28$
B $28 > 25 > 18$
C $28 < 25 <1 8$
D $18 > 25 > 28$

5. Start at the number 20. Skip count by 2s. What is the eighth number you will say?

Record your answer. Then fill in the bubbles. Be sure to use the correct place value.

⓪	⓪
①	①
②	②
③	③
④	④
⑤	⑤
⑥	⑥
⑦	⑦
⑧	⑧
⑨	⑨

6. Erin is wants to paste 36 tiles in equal rows. There will be 4 tiles in each row. How many rows of tiles will there be? (Hint: *Make a model or act out.*)

7. There are 12 boys and 15 girls in a class. The teacher wants to form groups with the same number of boys and the same number of girls in each group. What is the greatest number of groups she can make with all of the children? Explain your answer. (Hint: *Predict and test.*)

8. Juice boxes come in packages of 3. String cheese comes in packages of 5. Yvette wants the same number of each. What is the fewest number of each she can buy? (Hint: *Use logical reasoning.*)

9. Herb will buy bags of oranges. José will buy bunches of bananas. They will each buy the same number of fruits.

What is the least number of oranges or bananas they can buy?

10. Kevin ate 2 bags of fruit snacks each day at lunch from Monday to Friday. At the end of the day on Friday, he had 3 bags of fruit snacks left. How many bags of fruit snacks did Kevin have on Monday?

Strategies: Picture • Model/Act Out • Organized List • Table/Graph • Pattern • Predict & Test
Logical Thinking • Work Backward • Simpler Problem • Number Sentence

15 Fractions and Decimals

Fractions and decimals name parts of a whole.
Decimals name tenths, hundredths, and thousandths.

Example 1

Three children each planted 24 daffodils. Of these, $\frac{2}{3}$ of Kareem's bloomed, $\frac{1}{2}$ of Joe's bloomed, and $\frac{4}{8}$ of Carlos's bloomed. Which two children have the same number of blooming flowers? How many of their flowers bloomed?

Find Out First you need to find equivalent fractions. Then you can find out how many flowers bloomed.

Plan You can *make a model or act out* to solve this problem.

Solve First, use fraction bars to compare fractions.

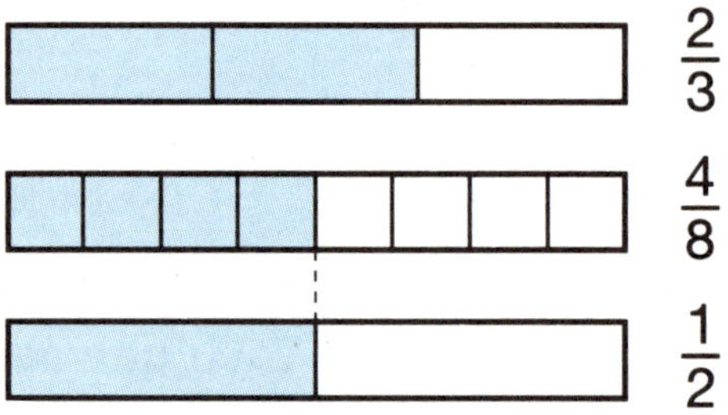

$\frac{4}{8}$ and $\frac{1}{2}$ are equivalent fractions.

Then use counters to find how many parts of 24 the equivalent fractions stand for.

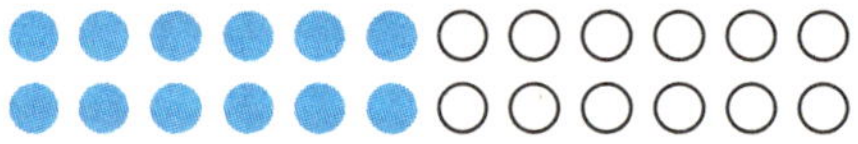

Show 2 equal groups.

$\frac{1}{2}$ of 24 is 12.

Answer: Joe and Carlos have the same number of bloomed flowers. Each had 12 blooming flowers.

Look Back You can draw a picture to check your answer.

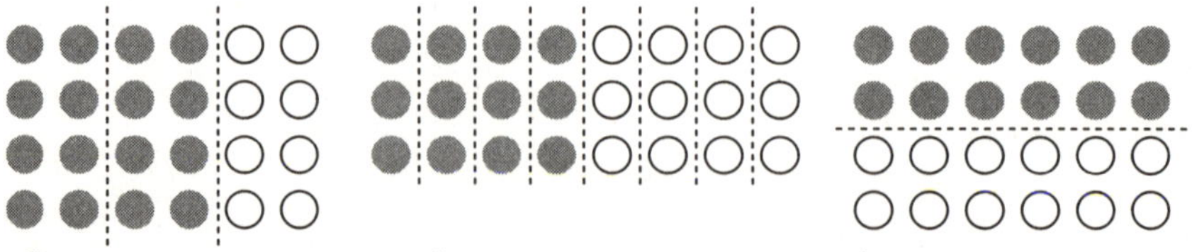

$\frac{2}{3}$ of 24 is 16 ✓ $\frac{4}{8}$ of 24 is 12 ✓ $\frac{1}{2}$ of 24 is 12 ✓

Try It

Example 2

Chang wrote how much snow and rain 3 cities get in January on 3 cards. He wrote the city names on 3 other cards. Then he dropped the cards.

New York | Baltimore | 3.6 inches

Houston | 3.7 inches | 3.5 inches

Find Out

Use these clues to help Chang match up the cards:

- New York has more snow and rain than Baltimore.
- Houston has more than New York.

How many inches of snow and rain does each city get?

You need to find out how much snow and rain each city gets so that you can match up the cards.

Plan

You can *use logical thinking* to solve the problem.

Solve

Write the information from each clue with the > symbol.

Houston > New York   New York > Baltimore

Use logical thinking to combine the clues.

Houston > New York _____ Baltimore.

Use the > symbol to order the decimals.

3.7 > 3.6 > _____

Match the cities with the decimals.

Answer: In January, Houston gets 3.7 inches, New York gets _____ inches and __________ gets _____ inches.

Look Back

How else can you solve this problem? Show your work.

____

Practice Test Questions

Find Out • Plan • Solve • Look Back

1. Billy is making a cake. He needs $\frac{1}{4}$ teaspoon of sugar and $\frac{1}{2}$ cup of flour. Which amounts are equivalent to these ingredients? (Hint: *Make a model or act out.*)

A $\frac{2}{8}$ teaspoon sugar, $\frac{3}{8}$ cup flour
B $\frac{2}{8}$ teaspoon sugar, $\frac{4}{8}$ cup flour
C $\frac{3}{8}$ teaspoon sugar, $\frac{3}{4}$ cup flour
D $\frac{4}{8}$ teaspoon sugar, $\frac{2}{8}$ cup flour

2. Matt, Keesha, and Ellen each ate part of their candy bars. One of them ate $\frac{1}{5}$, another ate $\frac{2}{3}$, and the other ate $\frac{3}{4}$. Matt ate more than Ellen. Matt ate less than Keesha. Which sentence below is **true**? (Hint: *Use logical thinking.*)

A Matt ate $\frac{2}{3}$ of his candy bar.
B Keesha ate $\frac{1}{5}$ of her candy bar.
C Ellen ate $\frac{3}{4}$ of her candy bar.
D Ellen ate $\frac{2}{3}$ of her candy bar.

3.

Sal's Sandwich Shop	
Chicken	$4.25
Cheese	$3.95
Tuna	$4.00

Which of the following shows the sandwiches in order from least to greatest price?

A tuna, chicken, cheese
B chicken, tuna, cheese
C cheese, tuna, chicken
D chicken, cheese, tuna

4. What is the greatest number that has:

- a digit in the ones, tenths, and hundredths places, and
- only digits 8 and 9?

Record your answer. Then fill in the bubbles. Be sure to use the correct place value.

	.		
⓪		⓪	⓪
①		①	①
②		②	②
③		③	③
④		④	④
⑤		⑤	⑤
⑥		⑥	⑥
⑦		⑦	⑦
⑧		⑧	⑧
⑨		⑨	⑨

5. Use the table to compare the times of the runners.

Summer Olympics Champions
Women's 200-Meter Race

Year	Runner	Time
1988	Griffith-Joyner	21.34 seconds
1992	Torrence	21.81 seconds
1996	Perec	22.12 seconds
2000	Jones	21.84 seconds

List the winning times in order from fastest time to slowest. Who ran the fastest? (Hint: *Make an organized list.*)

6. Alice is painting a wall in her room. She will paint $\frac{1}{8}$ of the wall pink, $\frac{1}{4}$ of the wall blue, and $\frac{5}{8}$ of the wall green. Will she paint more of the wall pink or blue? Explain how you know. (Hint: *Make a model or act out.*)

____

____

____

____

7. For lunch, Angie drank $\frac{1}{2}$ cup milk. Paulo drank $\frac{3}{4}$ cup apple juice, and Greg drank $\frac{3}{6}$ cup water. Which two children had the same amount to drink for lunch? Explain your answer.

____

____

____

____

8. Every day, Mel jogs 1.2 miles. Every day, Sara jogs 2.4 miles. Who jogs more miles in a week? Explain how you know.

____

____

____

____

9. Katy has three kittens:
Scruffy weighs 2.75 pounds.
Fluffy weighs 2.69 pounds.
Muffy weighs 2.7 pounds.

List the kittens in order from lightest to heaviest.

Strategies: Picture • Model/Act Out • Organized List • Table/Graph • Pattern • Predict & Test • Logical Thinking • Work Backward • Simpler Problem • Number Sentence

16 Add and Subtract Like Fractions

You can add or subtract fractions that have the same denominators. Add or subtract the numerators. Write the denominator.

Example 1

Hector made a list of what he ate at his meals on Monday.

Breakfast: Cereal, 1 cup; Fruit, $\frac{1}{2}$ cup

Lunch: Meat, $\frac{3}{8}$ pound; Vegetables, $\frac{1}{2}$ cup

Dinner: Fish, $\frac{4}{8}$ pound; Vegetables, 1 cup

How much meat or fish did Hector eat on Monday?

Find Out

You want to find out how much meat or fish Hector eats on Monday. The list gives too much information.

Plan

You can *write a number sentence* to solve this problem.

Think: *What operation will you choose?*

You can use addition because you need to find a total. Hector ate $\frac{3}{8}$ pound of meat at lunch. He ate $\frac{4}{8}$ pound of fish at dinner.

Solve

$\frac{3}{8} + \frac{4}{8} = \square$ *Write a number sentence.*

$\frac{3}{8} + \frac{4}{8} = \frac{7}{8}$ *Add the numerators.*
Write the denominator.

Answer: Hector ate $\frac{7}{8}$ pound of meat or fish on Monday.

Look Back

You can make a model to check your answer.

Use fraction bars for $\frac{3}{8}$ and $\frac{4}{8}$.

$\frac{3}{8} + \frac{4}{8} = \frac{7}{8}$ ✔

Try It

Example 2

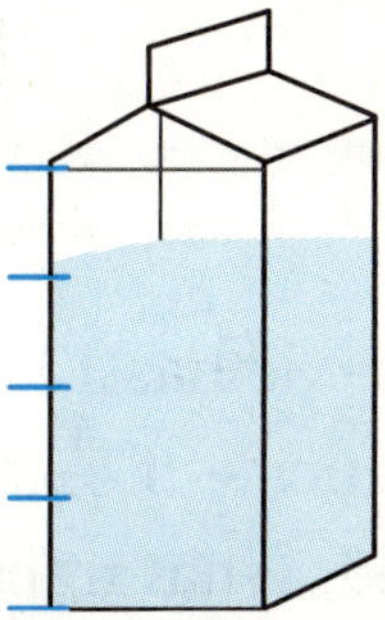

Nick wanted some milk. The milk container was $\frac{3}{4}$ full.

Nick drank $\frac{1}{4}$ of the container. How much of the quart of milk is left?

Find Out

You know:
- There was ________ quart of milk.
- Nick drank ________ quart of milk.

You need to find out how much of the quart is left.

Plan

You can *draw a picture* to solve the problem.

Solve

Draw a fraction strip to show fourths.

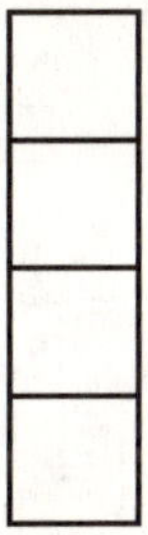

$\frac{3}{4}$

Show the number of fourths Nick started with.

Cross out the number of fourths Nick drank.

Use the picture to tell how many fourths are left.

Answer: There is ______ quart of milk left.

Look Back

How can you solve this problem another way?

____

Practice Test Questions

Find Out • Plan • Solve • Look Back

1. Maxie has 12 red and blue pencils. If $\frac{4}{12}$ of the pencils is red, what fraction of her pencils is blue? (Hint: *Write a number sentence.*)

A $\frac{1}{12}$ **C** $\frac{7}{12}$

B $\frac{5}{12}$ **D** $\frac{8}{12}$

2. There is $\frac{7}{8}$ of an apple pie on the table. David ate $\frac{3}{8}$ of the pie. Which number sentence could you use to find the part of the pie that David did **not** eat? (Hint: *Draw a picture.*)

A $\frac{7}{8} + \square = \frac{3}{8}$ **C** $\frac{7}{8} + \frac{3}{8} = \square$

B $\frac{7}{8} - \square = \frac{3}{8}$ **D** $\square - \frac{3}{8} = \frac{7}{8}$

3. Rene made 3 glasses of chocolate milk. He used $\frac{1}{8}$ cup chocolate syrup in each glass. Now there is $\frac{2}{8}$ cup syrup left. How much syrup was in the jar to start with? (Hint: *Work backward.*)

A $\frac{3}{8}$ cup **C** $\frac{5}{8}$ cup

B $\frac{4}{8}$ cup **D** $\frac{6}{8}$ cup

4. Look at the fractions and symbols below.

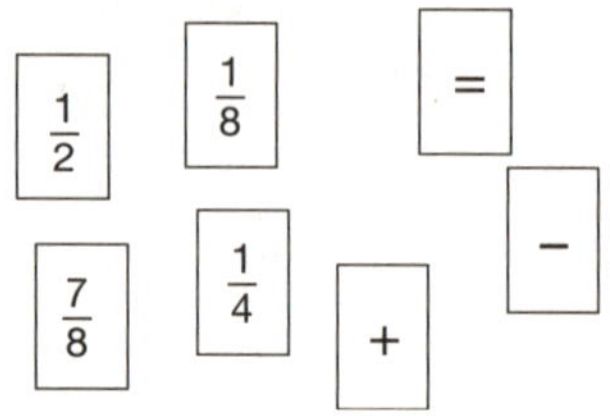

The cards are arranged to make number sentences. Which one is **not** a true number sentence?

A $\frac{7}{8} + \frac{1}{8} = \square - \frac{1}{4}$

B $\frac{7}{8} - \frac{1}{8} = \square + \frac{1}{4}$

C $\frac{7}{8} - \frac{1}{2} = \frac{1}{4} + \frac{1}{8}$

D $\frac{7}{8} - \square = \frac{1}{8} + \frac{1}{2}$

5. What number belongs in the box to make this number sentence true?

$$\frac{\square}{12} - \frac{4}{12} = \frac{6}{12}$$

Record your answer. Then fill in the bubbles. Be sure to use the correct place value.

⓪	⓪
①	①
②	②
③	③
④	④
⑤	⑤
⑥	⑥
⑦	⑦
⑧	⑧
⑨	⑨

6. The table shows how much each piece of fruit weighs.

Weights of Fruits

Apple	$\frac{5}{16}$ pound
Peach	$\frac{3}{16}$ pound
Banana	$\frac{4}{16}$ pound
Pear	$\frac{3}{16}$ pound

Vera cut up some pieces of fruit for a fruit salad. She used $\frac{9}{16}$ pound of fruit in it. Which pieces of fruit did Vera cut up? (Hint: *Predict and test.*)

7. The circle graph shows how Dan spent his birthday money.

How Dan Spent His Money

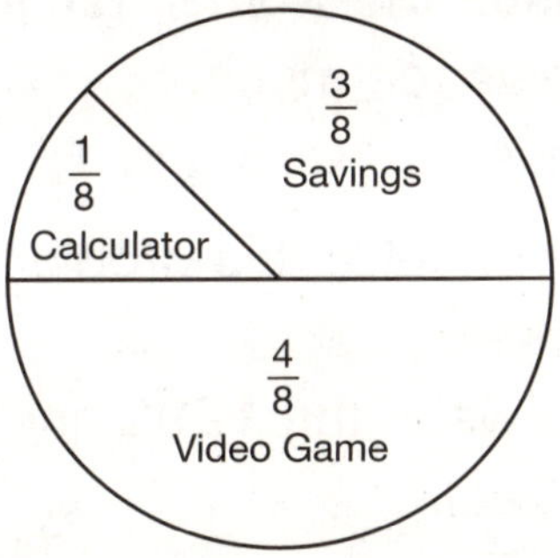

What fraction of his money did Dan use to buy things? (Hint: *Write a number sentence.*)

8. A tomato plant is $\frac{1}{6}$ foot taller than a pepper plant. The sum of the heights of the two plants is $\frac{5}{6}$ foot. How tall is the pepper plant? Show your work.

9. Leroy wants to buy some cashew nuts.

Nuts For Sale	
$\frac{1}{4}$ pound	
Cashews	\$2.00
Peanuts	\$0.50
Almonds	\$1.00

How many pounds of cashews can he buy for \$8? Explain how you know.

____

____

____

Strategies: Picture • Model/Act Out • Organized List • Table/Graph • Pattern • Predict & Test • Logical Thinking • Work Backward • Simpler Problem • Number Sentence

17 Add and Subtract Decimals

To add or subtract decimals, line up the decimal points. Then add or subtract the same way you do for whole numbers.

Example 1

At the Summer Olympics, the United States won the Women's 400-meter relay race four times in a row.

U.S. Women's 400-Meter Relay Race

Year	Winning Time
1984	41.65 seconds
1988	41.96 seconds
1992	42.11 seconds
1996	41.95 seconds

How much faster did the women run in 1988 than in 1992?

Find Out You want to find out how much faster the women ran in 1988 than in 1992. The table shows too much information. You need only the times for 1988 and 1992.

Plan You can *write a number sentence* to solve this problem.
Think: *What operation will you use?*
You need to find the difference between the 1992 and the 1988 winning times. Use subtraction.

Solve $42.11 - 41.96 = \square$
$\square = 0.15$
The women ran 0.15 second faster.

Look Back You can estimate to be sure your answer is reasonable.
41.95 and 42.11 are each about 42.
$42 - 42 = 0$, which is close to 0.15 ✔
0.15 second < 1.00 second ✔

Try It

Example 2

The graph shows the cost of a slice of pizza with 1 to 3 toppings at Joe's Pizza Palace.

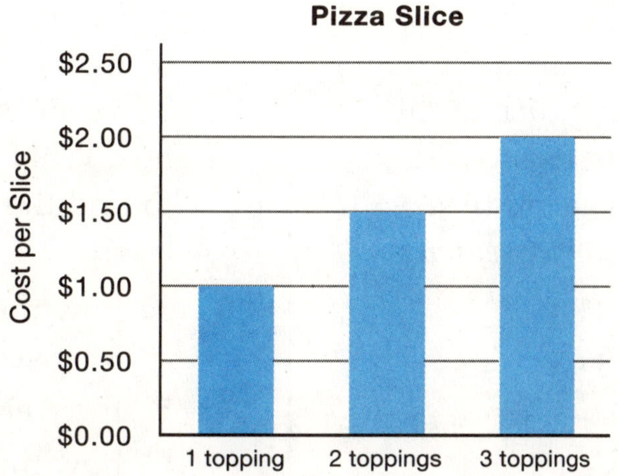

How much does Joe charge for each topping on a slice of pizza?

Find Out You know: ________ = the cost of 1 slice with 1 topping.
________ = the cost of 1 slice with 2 toppings.
________ = the cost of 1 slice with 3 toppings.

You need to find out how much each topping costs.

Plan You can *look for a pattern* to solve the problem. Then you can find out how the price changes for each new topping.

Solve $1.50 – _____ = _____ *Find the difference between 1 topping and 2 toppings.*

$2.00 – _____ = _____ *Find the difference between 2 toppings and 3 toppings.*

For each topping, the cost goes up by _________.

Answer: Joe charges _____ for each topping.

Look Back How can you solve this problem another way?

____

Practice Test Questions

Find Out • Plan • Solve • Look Back

1. A greyhound can run up to 39.35 miles per hour. This is 11.46 miles per hour faster than a human can run. Which number sentence can you use to find how fast a human can run? (Hint: *Write a number sentence.*)

A 39.35 + □ = 11.46

B □ – 11.46 = 39.35

C 39.35 = 11.46 + □

D 39.35 + 11.46 = □

2. Mi Sook makes muffins every Sunday morning. The table shows how much flour is left after she makes the muffins each week.

Week	1	2	3	4
Leftover Flour	16.5 ounces	12.5 ounces	? ounces	4.5 ounces

Which number sentence shows how to find the missing amount in week 3? (Hint: *Look for a pattern.*)

A 12.5 + 4

B 4.5 – 4

C 12.5 – 4

D 4.5 – 4

3. A small box of raisins weighs 1.75 ounces less than a medium box. The large box weighs 3.75 ounces more than the small box. What do you need to know to find the weight of a medium box?

A how much more a medium box weighs than a small box

B the weight of a small box

C how much less a medium box weighs than a large box

D You already have enough information.

4. A crate weighs 9.4 pounds. Tom fills it with 39.6 pounds of oranges. How many pounds do the crate and the oranges weigh?

Record your answer. Then fill in the bubbles. Be sure to use the correct place value.

⓪	⓪
①	①
②	②
③	③
④	④
⑤	⑤
⑥	⑥
⑦	⑦
⑧	⑧
⑨	⑨

5. The door to the classroom is 2.5 feet taller than the window above it. The sum of the heights of the window and door is 15 feet. What is the height of the window? (Hint: *Try the predict and test strategy.*)

6. The table shows the money each zoo spent in 2002.

2002 Zoo Budgets

Zoo	Budget
Buffalo Zoo	$4.8 million
Memphis Zoo	$8.0 million
Miami Metrozoo	$8.4 million
Cleveland Metroparks Zoo	$12.1 million

The San Francisco Zoo spent $1 million more than the Toledo Zoo. The Toledo Zoo spent $2.2 million more than Buffalo and Memphis zoos together. How much did the San Francisco Zoo spend in 2002? (Hint: *Write a number sentence.*)

7. The table shows the normal amount of rain and snow in Houston, Texas, from January through April.

Rain and Snow in Houston, Texas

Month	Normal Amount (in inches)
January	3.7
February	3.0
March	3.4
April	3.6

In May, the normal amount is 2.2 inches more than the least amount in the table. What is the normal amount in May? Show how you got your answer.

8. Renee needs 3 pieces of ribbon for an art project. The pieces must be 1.50 feet, 1.20 feet, and 1.75 feet long. She has a roll of ribbon that is 5 feet. How much ribbon will be left on the roll?

Strategies: Picture • Model/Act Out • Organized List • Table/Graph • Pattern • Predict & Test • Logical Thinking • Work Backward • Simpler Problem • Number Sentence

18 Metric and Customary Measurement

Customary measurements are units such as inches, feet, pounds, quarts, and gallons. Metric measurements are units such as meters, grams, and liters.

Example 1

Rosa is 13 years old. She is 64 inches tall. She grew 2 inches a year for each of the last 3 years. How tall was Rosa when she was 10 years old?

Find Out

You know that Rosa grew 2 inches each year for the last 3 years. You know that now she is 64 inches tall and 13 years old.

You want to find Rosa's height when she was 10 years old.

Plan

You can *make a table* to find out how tall she was when she was 12 years old, then 11 years old, and then 10 years old.

Solve

Make a table. Start with 64 inches at 13 years old. Continue until you reach 10 years old.

Age	13	12	11	10
Height (in inches)	64	62	60	58

Read the information from the table.

Answer: Rosa was 58 inches tall.

Look Back

You can work backward to be sure that your answer is correct.
At 13, she was 64 inches tall.
At 12, she was 64 – 2 inches = 62 inches.
At 11, she was 62 – 2 inches = 60 inches.

At 10, she was 60 – 2 inches = 58 inches. ✔
The answer is correct.

Try It

Example 2

You want to measure exactly 10 ounces of milk into a glass. You can measure with these containers.

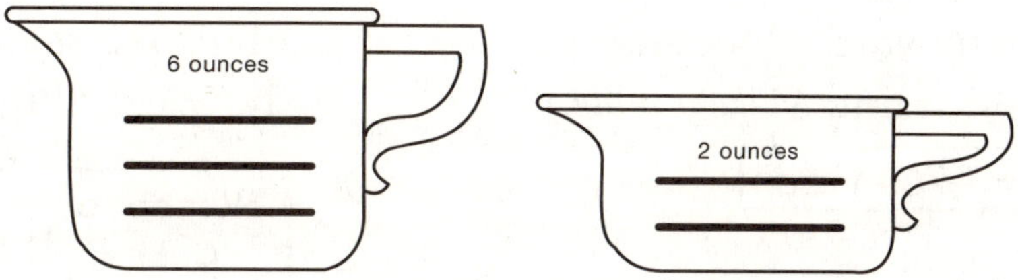

Find Out

You know:
- One container holds 6 ounces.
- The other container holds 2 ounces.

You need to find out how to measure exactly 10 ounces of milk into a glass.

Plan

You can *use logical thinking* to solve the problem.
Think: *How many times will you have to fill each container to have 10 ounces of milk?*

Solve

1. Fill the container that holds 6 ounces, and pour the milk into the glass.
   Now you have ______ounces of milk in the glass.
2. Fill the container that holds 2 ounces, and pour it into the glass.
   Now you have ______ounces of milk in the glass.
3. Fill the container that holds 2 ounces, and pour it into the glass.
   Now you have ______ounces of milk in the glass.

Answer: To measure exactly 10 ounces of milk, you can

____

____

Look Back

How can you solve this problem another way?

____

Practice Test Questions

Find Out • Plan • Solve • Look Back

1. Aisha runs 2 miles every day. How many yards does she run each day? (Hint: *Make a table.*)

1 mile = 1,760 yards

A 880 yards
B 2,760 yards
C 2,820 yards
D 3,520 yards

2. Miguel caught a fish that is 4 inches longer than the fish Luis caught. The sum of the lengths of both fish is 1 yard. How long is the fish Miguel caught? (Hint: *Predict and test.*)

1 yard = 36 inches

A 16 inches
B 18 inches
C 20 inches
D 32 inches

3. Ted grew a watermelon that is 1 foot 7 inches long. What is the length of the watermelon in inches? (Hint: *Make a table.*)

A 17 inches
B 19 inches
C 21 inches
D 23 inches

4. Raffi is going to bake cakes for the bake sale. The table shows how much milk he will need.

Number of Cakes	Milk (in cups)	Milk (in quarts)
1	4	1
2	8	2
3	12	?
4	16	4

Which expression shows how to find how many quarts of milk he needs for 3 cakes?

A $12 \div 4$
B $12 - 4$
C $12 \div 3$
D $12 - 3$

5. How many inches are there in 2 yards?

1 foot = 12 inches
1 yard = 3 feet

Record your answer. Then fill in the bubbles. Be sure to use the correct place value.

⓪	⓪
①	①
②	②
③	③
④	④
⑤	⑤
⑥	⑥
⑦	⑦
⑧	⑧
⑨	⑨

6. One glass has 4 more ounces of juice than another. You pour the juice from both glasses into a pitcher. You have 32 ounces of juice. How much juice was in each glass? (Hint: *Use logical thinking*.)

7. Peggy picked 3 pounds of plums at the orchard. Sally picked 40 ounces of plums. Who picked the heavier basket of plums? (Hint: *Make a table.*)

1 pound = 16 ounces

8. Barry bought a sheet of plywood. He used it to make a bookshelf. He used 8 feet to make the shelves. He used 6 feet for the sides. He has 7 feet of the plywood left. How long was the sheet of plywood? (Hint: *Work backward.*)

9. Everett had 9 feet of fencing. He cut it into 4 equal sections to put around a garden. He had 1 foot of fencing left. Explain how to find the length of each section of fence around the garden.

____

____

____

____

10. Mark has 4 puppies.

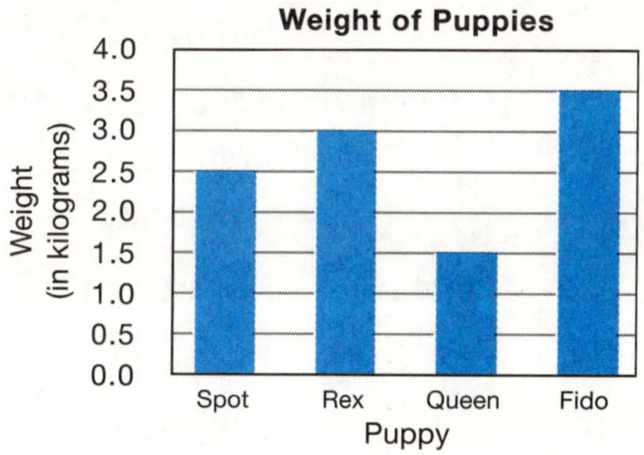

How much will Rex weigh if he gains 3 kilograms?

19 Time and Temperature

Clocks measure time in seconds, minutes, hours, and days. Thermometers measure temperatures in Fahrenheit (F) degrees or Celsius (C) degrees.

Example 1

The Rodriguez family is going to see a movie. The clocks show the starting and ending times for the movie. How long is the movie?

Starting Time

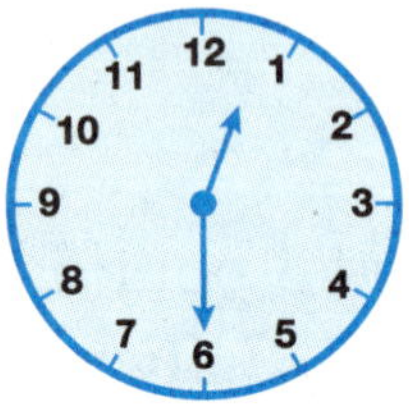

Ending Time

Find Out You need to find out how much time passes from 12:30 to 2:00.

Plan You can *act out* this problem. Use a model of a clock. First count how many hours pass and then how many minutes.

Solve Start at 12:30.

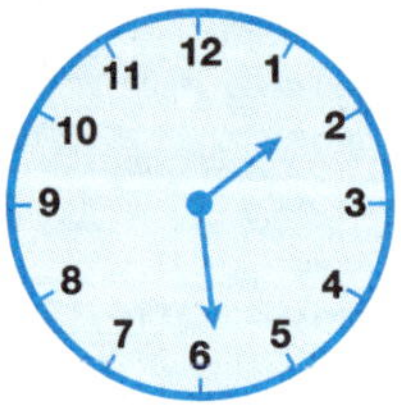

Move the hands to see how many full hours pass. 1 hour

Then move the hands to see how many more minutes pass until you reach 2 o'clock. 30 minutes

Answer: The movie is 1 hour and 30 minutes long.

Look Back You can make an organized list to check your answer. List the hours and then the minutes:

12:30 to 1:30 = 1 hour
1:30 to 2:00 = 30 minutes
1 hour + 30 minutes = 1 hour and 30 minutes ✔

Try It

Example 2

It is snowing today. Alan has no school today. He looks at the Fahrenheit thermometer below.

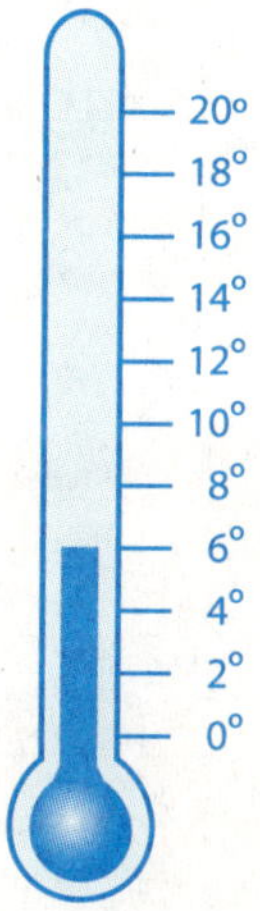

The weather report says the temperature should go up another 10 degrees today. What will be the high temperature for today?

Find Out You know:

- The temperature now is 6° Fahrenheit
- The temperature will go up 10 degrees

You need to find out what the temperature will be.

Plan You can *look for a pattern* to solve the problem. You can find out what the temperature will be after it goes up by 2° and then by 2° again.

Solve Continue the pattern.

6°F 8°F __°F __°F __°F __°F

+ 2° + 2° + 2° + 2° + 2° = 10° more

Answer: Today the high temperature will be _____° Fahrenheit.

Look Back How can you solve this problem another way?

____

Practice Test Questions

Find Out • Plan • Solve • Look Back

1. Chef Collins cooks his chicken soup for 2 hours and 15 minutes. He started cooking the soup at:

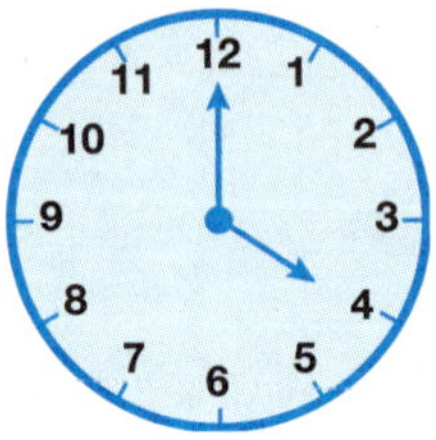

At what time will the chicken soup be ready? (Hint: *Make a model or act out.*)

A 6:00 **C** 6:30
B 6:15 **D** 6:45

2. Abe visits his aunt every 5 days. He visits her on April 2.

April

SUN	MON	TUE	WED	THU	FRI	SAT
		1	2	3	4	5
6	7	8	9	10	11	12
13	14	15	16	17	18	19
20	21	22	23	24	25	26
27	28	29	30			

On which Tuesday this month will he visit his aunt? (Hint: *Look for a pattern.*)

A April 8 **C** April 22
B April 15 **D** April 29

3. Jan's school is in session from September 1 to June 30. What fraction of the year is she **not** in school?

1 year = 12 months

A $\frac{1}{6}$ **C** $\frac{1}{12}$
B $\frac{5}{6}$ **D** $\frac{5}{12}$

4. Look at the time line of the first four presidents of the United States.

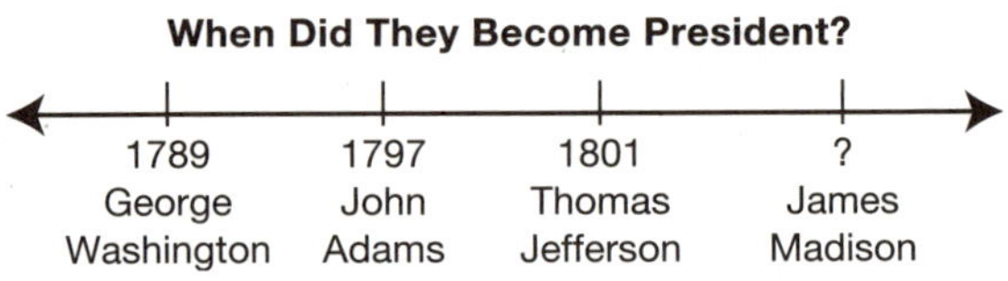

James Madison became president 20 years after the first president took office. What was that year?

Record your answer. Then fill in the bubbles. Be sure to use the correct place value.

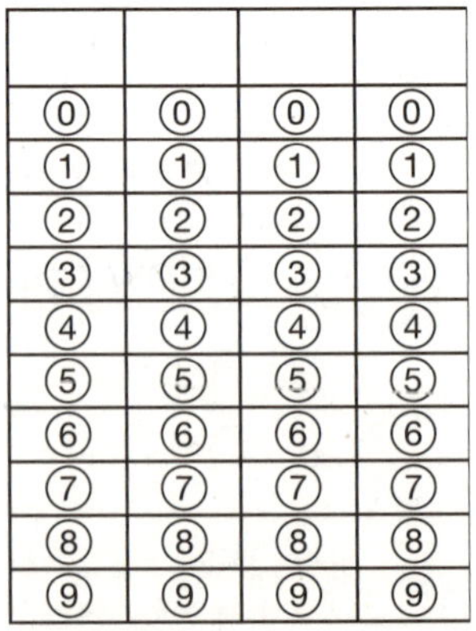

5. Marta read the temperature when she went outside. It was 25° Fahrenheit. She came inside when the temperature dropped 8°. What is the temperature now? (Hint: *Draw a diagram.*)

6. The Wu family returned from vacation on June 24. They spent a week at the beach and another week camping. On what date did the Wu's vacation begin? (Hint: *Work backward.*)

1 week = 7 days

7. Denny ran around the track in 80 seconds. It took Max 1 minute and 10 seconds to run the same distance. Who ran the fastest? (Hint: *Make a table.*)

1 minute = 60 seconds

8. Jack wants to be in New York City by 3:00 P.M. Which bus should he take? At what time will he arrive in New York City?

To New York City		
Bus Number	**Departs**	**Arrives**
416	6:30 A.M.	11:45 A.M.
395	9:00 A.M.	2:15 P.M.
503	1:15 P.M.	6:30 P.M.
162	4:30 P.M.	9:45 P.M.

9. Today it takes Mario 1 hour to have breakfast and shower. He spends $2\frac{1}{2}$ hours doing his homework. Then he visits his friends for 2 hours, until 3:00 P.M. Make a schedule for Mario. List each activity and the time.

____

____

____

____

Strategies: Picture • Model/Act Out • Organized List • Table/Graph • Pattern • Predict & Test • Logical Thinking • Work Backward • Simpler Problem • Number Sentence

20 Plane Figures

You can name a polygon by the number of sides it has. The perimeter is the distance around the polygon. The area is the number of square units needed to cover the polygon.

Example 1

Mr. Yumi is writing a math test for his students.

Math Test

Name each figure.

Mr. Yumi is following a pattern. What kind of figure will he show for question 6?

Find Out You need to find out what figure will be shown for question 6.

Plan You can *look for a pattern* to solve this problem.

Solve Count the number of sides for each figure.

Math Test

Name each figure.

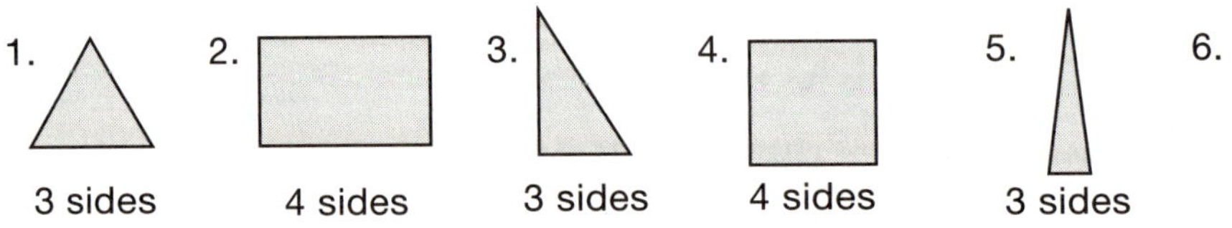

Describe the pattern.
3 sides followed by 4 sides

Follow the pattern.
Since the figure for question 5 has 3 sides, the figure for question 6 will have 4 sides.

Answer: Mr. Yumi will show a quadrilateral for question 6.

Look Back Check to see that the answer continues the pattern.
The pattern is 3 sides followed by 4 sides. If Mr. Yumi draws a quadrilateral for question 6, the pattern continues. ✔

Try It

Example 2

Alda painted a frame around her design.

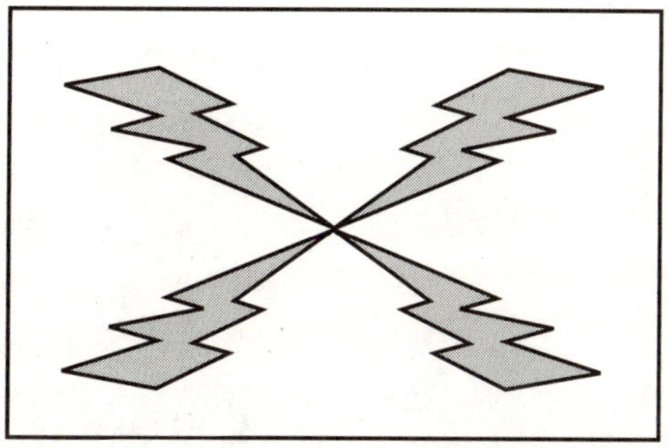

What is the perimeter of her frame?

Find Out

You know:
- The frame is a rectangle.
- The perimeter is the distance around the frame.

You need to find the perimeter of the frame.

Plan

You can *act out* the problem. Use a centimeter ruler to measure Alda's frame.

Solve

Measure the distance around the rectangle with a ruler.

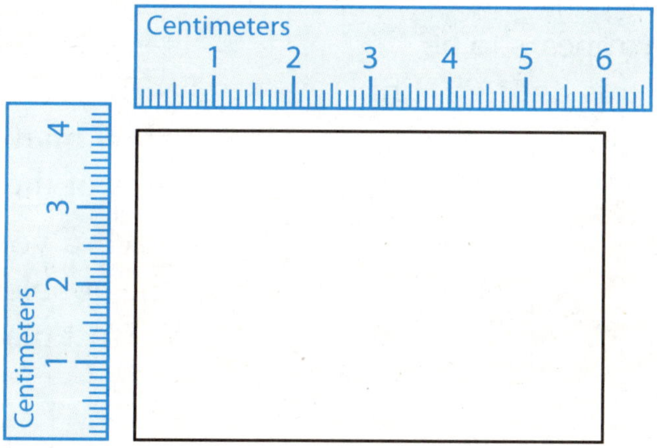

Answer: The perimeter of the frame is ___________ centimeters.

Look Back

How else can you solve this problem? Show your work.

____

Practice Test Questions

Find Out • Plan • Solve • Look Back

1. 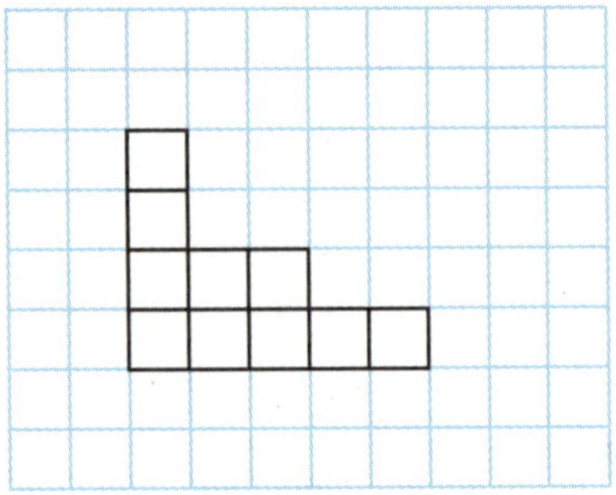

Which of the following could you use to find the perimeter of the figure shown above? (Hint: *Make a model or act out.*)

A 4 + 2 + 2 + 2

B 2 + 2 + 1 + 4 + 5

C 5 + 3 + 2

D 1 + 2 + 2 + 1 + 2 + 1 + 5 + 4

2. Which figure appears to be congruent to this figure? (Hint: *Draw a picture.*)

A 

C 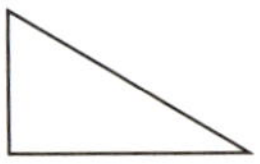

B 

D 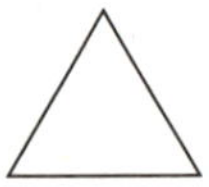

3. Which is **not** true about the shape shown below?

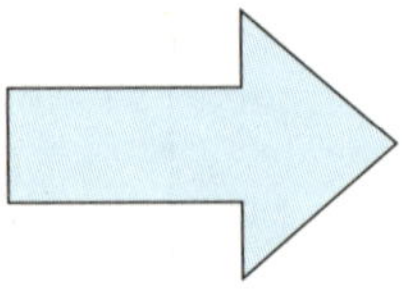

A It is a triangle.

B All the sides are not the same lengths.

C It is not a quadrilateral.

D It is symmetrical.

4. This is the floor in Joey's classroom.

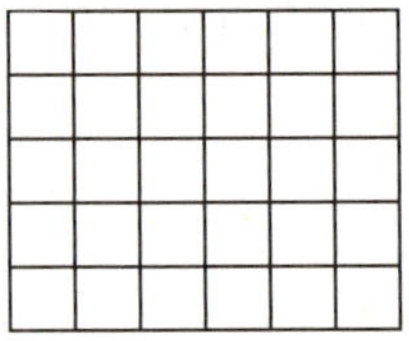

How many square tiles cover the area of the floor?

Record your answer. Then fill in the bubbles. Be sure to use the correct place value.

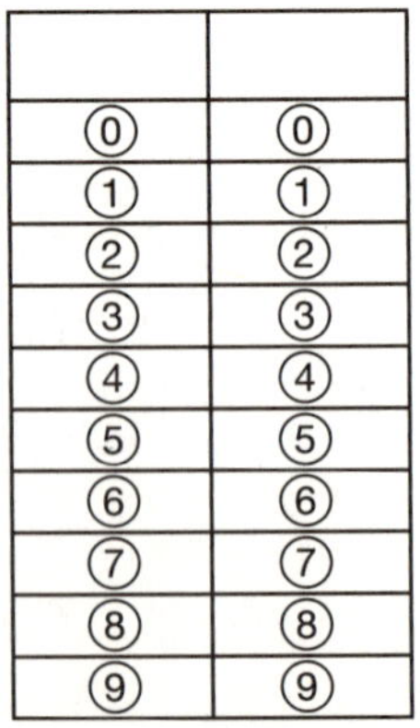

5. The art teacher drew a pattern along the top border of her bulletin board. She used repeating shapes. Use the clues to draw the pattern. (Hint: *Look for a pattern.*)

- The pattern has a triangle, a rectangle, and a square.
- The triangle is not first. The rectangle is not last.
- The square is in the middle.

6. How many symmetrical shapes can you count in this picture? List them. (Hint: *Make a model or act out.*)

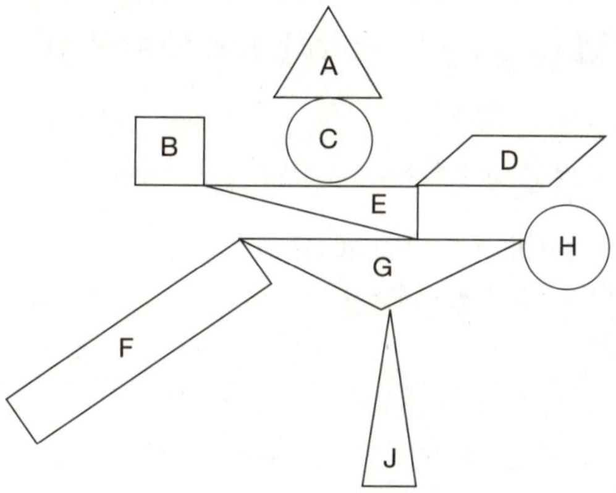

7. Find the perimeter and the area of the 2 figures below. Do they have the same perimeters? Do they have the same areas? Explain.

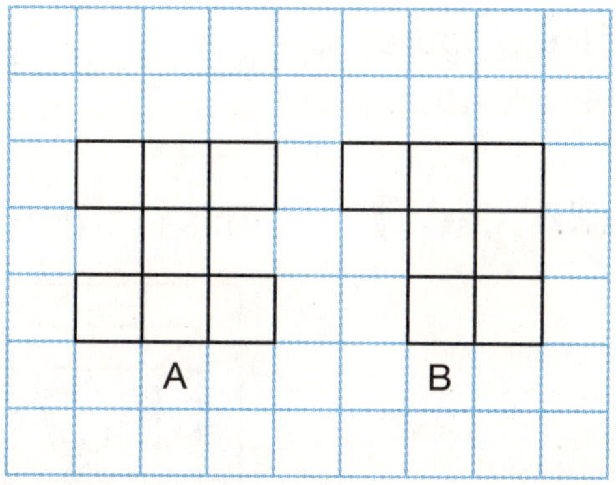

____

____

____

____

8. Tracy drew 2 polygons. The sum of their sides is 7. The difference in their sides is 1. What 2 polygons could Tracy have drawn?

Strategies: Picture • Model/Act Out • Organized List • Table/Graph • Pattern • Predict & Test Logical Thinking • Work Backward • Simpler Problem • Number Sentence

21 Solid Figures

You can name solid figures by the types and numbers of sides they have. The volume of a solid figure is the number of cubic units needed to fill the figure.

Example 1

This is a candleholder Teresa got as a gift.

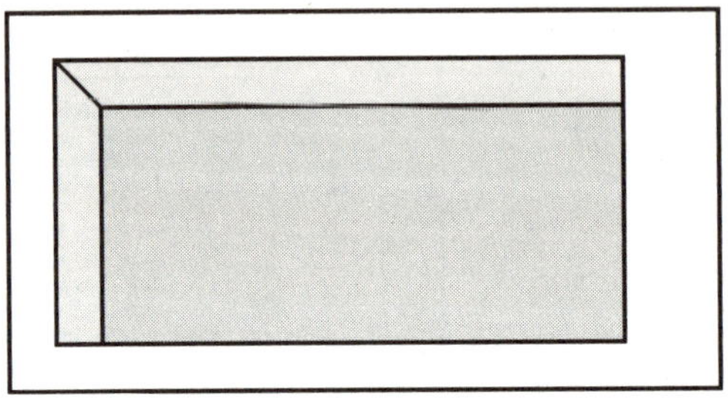

Now she wants to buy a candle that will fit into the candleholder. What are two solid shapes Teresa can choose for her candle?

Find Out Look at the shape of the candleholder. It is a rectangle. You need to find which solid figures have a base that is a rectangle.

Plan You can *make a model or act out* to solve this problem.

Solve

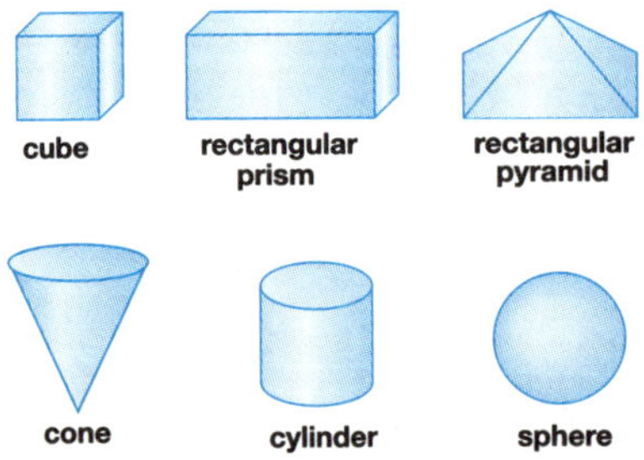

Make or look at models of solid shapes.

The rectangular prism and the rectangular pyramid both have rectangles as the base.

Answer: Teresa can buy a candle that is a rectangular prism or a rectangular pyramid to fit the candleholder.

Look Back You can trace the bases of the figures to see which ones are rectangles. ✔

Try It

Example 2

The picture shows how Tori arranged some blocks.

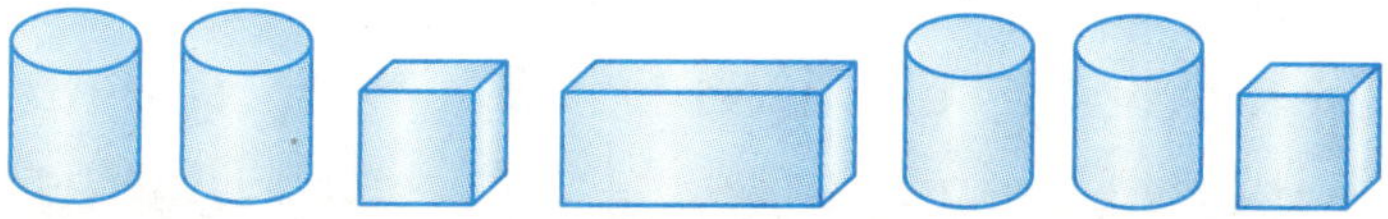

If Tori continues the pattern, what shape will the next block be?

Find Out

You know:
- The blocks are cylinders, cubes, and a rectangular prism.
- The blocks follow a pattern.

You need to find what block comes next.

Plan

You can *look for a pattern* to solve the problem. To do this you must first describe the pattern.

Solve

The pattern Tori followed is ______________________

followed by ______________________________________

followed by ______________________________________.

Then the pattern repeats up to the __________________.

Answer: The next block Tori will put on the shelf will be a

____________________.

Look Back

How can you solve this problem another way?

____

Practice Test Questions

Find Out • Plan • Solve • Look Back

1. Which two solid figures are used to make this figure? (Hint: *Make a model or act out.*)

A cylinder and cone
B sphere and cone
C square pyramid and cylinder
D cylinder and sphere

2. Dana wants to build a box like this cube.

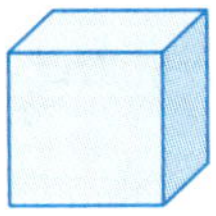

Which shows the pieces of wood she will need to build the box? (Hint: *Draw a picture.*)

A 

B 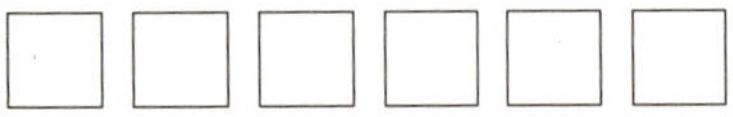

C 

D 

3. Which sentence is true for these two figures?

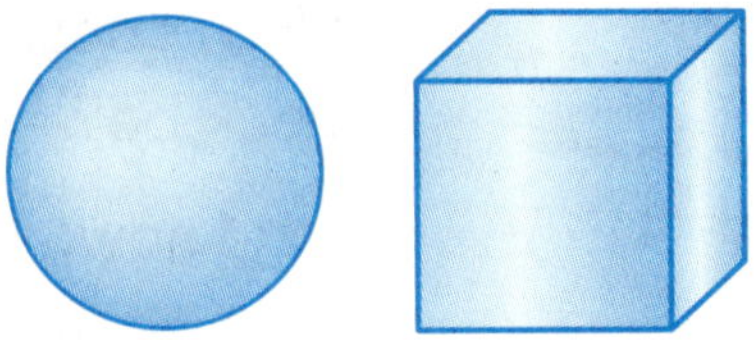

A Both have a curved surface.
B Neither has a curved surface.
C One has no flat faces.
D Both have flat faces.

4. How many faces do 2 rectangular pyramids have?

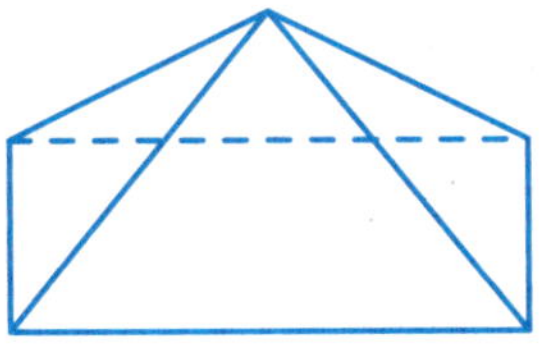

Record your answer. Then fill in the bubbles. Be sure to use the correct place value.

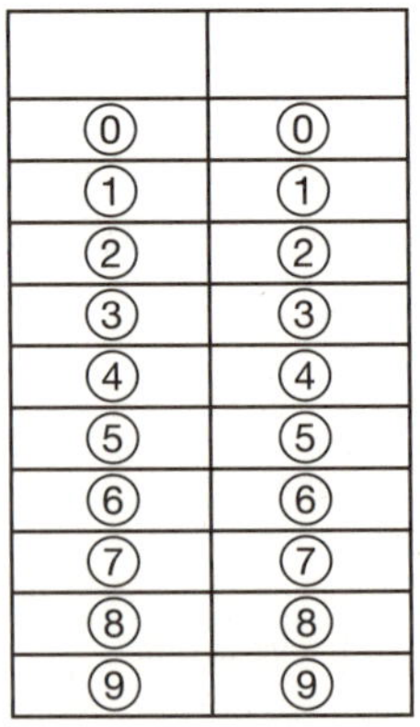

5. Ted decides to have a can of soup for lunch. How many flat faces does the can have? What shape are the faces? (Hint: *Make a model or act out.*)

6. Amy drops a box of blocks on the floor. The blocks are shaped like cylinders, cones, cubes, and prisms. Which blocks can roll when they hit the floor? Which cannot? Tell why. (Hint: *Act out.*)

____

____

____

____

7. The cake shown below served over 100 people at a wedding. What solid figures are used in the cake?

8. Manuel stores his toys in 3 boxes. Each box is a rectangular prism. What is the total number of edges?

9. The art teacher cut a large sponge to make a solid figure. He used 2 of the faces to make sponge prints. The 2 faces made the same circle print. What solid figure was the sponge?

Strategies: **Picture • Model/Act Out • Organized List • Table/Graph • Pattern • Predict & Test Logical Thinking • Work Backward • Simpler Problem • Number Sentence**

22 One-Step Equations

Sometimes you can write an expression or a number sentence that describes a problem. You can use a □ to stand for the missing number. Then you can solve to find the missing number.

Example 1

An animal shelter is collecting blankets to put in the cages. There are 4 kittens in a cage. There are 2 puppies in a cage. The shelter needs 3 blankets for each cage. What expression can you use to find out the number of blankets needed for any number of cages?

Find Out

You know the number of blankets needed for 1 cage. You need to find an expression to find out how many blankets are needed for any number of cages.

Plan

You can *make a table* and *look for a pattern* to solve this problem. This problem has too much information. You do not need to know how many kittens or puppies are in a cage.

Solve

Cages	1	2	3	4	← □
Blankets	3	6	9	12	← □ × 3

Use □ for the number of cages.

So, 3 × □ is the number of blankets needed for □ cages.

Answer: Use the expression 3 × □ to find the number of blankets needed for any number of cages.

Look Back

You can work backward to check your work.

Divide the number of blankets by the number of cages. Your answer should always be 3. ✔

Try It

Example 2

Jan and Carla both make friendship bracelets. They sell their bracelets at a flea market each week. This week, Carla sells 11 bracelets. That is 3 more than Jan sells. How many bracelets does Jan sell?

Find Out

You know:
- Carla sells 11 bracelets.
- Carla sells 3 more bracelets than Jan sells.

You need to find out how many bracelets Jan sells.

Plan

You can *draw a picture* and *write a number sentence* to solve the problem.

Solve

Carla **X X X X X X X X X X X**

Jan **X X X** □

Show Xs for the 11 bracelets.

Show Xs for the 3 bracelets.

3 + □ = 11 **Think**: *What number plus 3 equals 11?*

3 + _____ = 11

Answer: Jan sells _____ bracelets

Look Back

How can you solve this problem another way?

____

Practice Test Questions

Find Out • Plan • Solve • Look Back

1. Gene earns \$6 per hour for baby-sitting. What expression can you use to find how much he earns for any number of hours? (Hint: *Make a table.*)

A $6 + \square$

B $6 - \square$

C $6 \times \square$

D $6 \div \square$

2. The life span of a beaver is about 5 years. This is 7 years fewer than the life span of a deer. Which number sentence can you use to find the life span of a deer? (Hint: *Draw a picture.*)

A $7 + 5 = \square$

B $7 - \square = 5$

C $7 \times 5 = \square$

D $5 + \square = 7$

3. Every student in the third grade gets 4 textbooks. The third grade gets 100 textbooks in all. Use this number sentence to find the number of students in the third grade.

$$4 \times \square = 100$$

Which is the solution to the number sentence? (Hint: *Solve a simpler problem.*)

A $\square = 400$

B $\square = 104$

C $\square = 99$

D $\square = 25$

4. Trisha wants to buy a TV for \$100. She saves \$5 each week.

Which question can you answer by finding $\square$?

$$\square \times 5 = 100$$

A How much is the sales tax on the TV?

B How many weeks will it take Trisha to save enough money?

C How much does Trisha have to earn to buy the TV?

D Not here

5. Solve the number sentence. What is $\square$?

$$\square - 17 = 15$$

Record your answer. Then fill in the bubbles. Be sure to use the correct place value.

⓪	⓪
①	①
②	②
③	③
④	④
⑤	⑤
⑥	⑥
⑦	⑦
⑧	⑧
⑨	⑨

6. Which value for □ will make the number sentence true? (Hint: *Predict and test.*)

$$4 \times \square = 36$$

7. You put quarters in a cookie machine. The table shows how many cookies you get.

Quarters	Cookies
1	2
2	4
3	6
4	8

How many cookies will you get if you put □ quarters in the machine? Write an expression. Explain your answer. (Hint: *Look for a pattern.*)

____

____

____

____

8. Hank is 8 years older than Steve. Hank is 12 years old. Write a number sentence to find Steve's age. Solve your number sentence.

9. What is the missing number in the first column?

	+3
3	6
7	10
11	14
?	18
19	22

10. At a museum, visitors can use headphones to learn about the exhibits. At 11 A.M., there are 43 headphones on the shelf. One class is using 21 headphones. Another class is using 25 headphones. How many headphones did the museum start with?

Strategies: Picture • Model/Act Out • Organized List • Table/Graph • Pattern • Predict & Test • Logical Thinking • Work Backward • Simpler Problem • Number Sentence

23 Ordered Pairs on a Coordinate Grid

An ordered pair has two numbers. The first number tells how far to the right to go. The second number tells how far up to go.

Example 1

The map shows some places in Kelly's neighborhood. Each square unit stands for 1 square block.

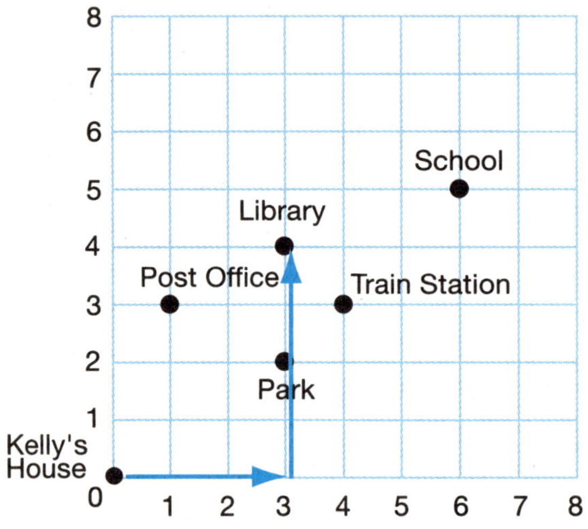

The library can be located on the map by the ordered pair (3,4). How many blocks from the park is the library?

Find Out You need to trace a path from Kelly's house to the library.

Plan You can *act out* this problem.

Solve The library is at (3,4)
start at (0,0)
Move 3 spaces to the right.
Then move up 4 spaces.

Answer: The map shows the library is 3 blocks to the right and 4 blocks up.

Look Back Work backward to check your answer.

Start at the library. Count 4 spaces down and 3 spaces left. Are you at Kelly's house? You are, so your answer is correct. ✔

Try It

Example 2

The map shows where some of the desks are in a third-grade classroom.

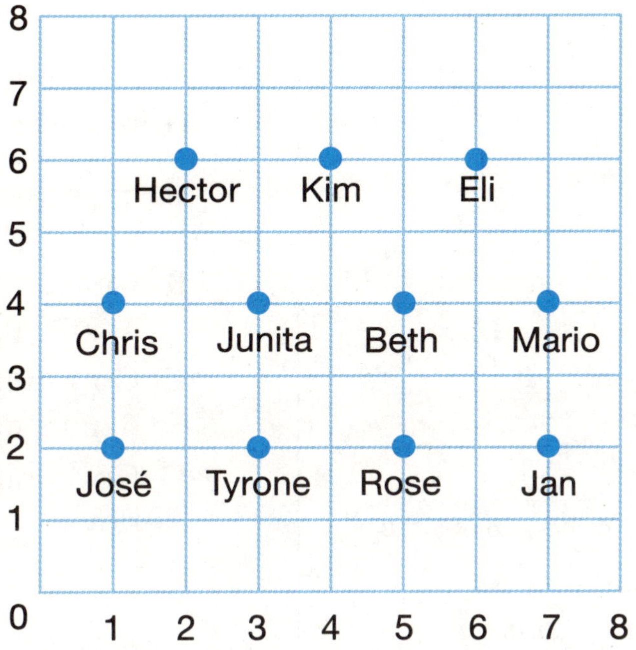

José walks to (5,2) and then to (5,4). Then he walks to the desk at (1,4) and then he returns to his own desk. Who sits at each desk he walks to? What shape does José's path make?

Find Out You need to find who sits at each desk José walks to. You need to find what shape his path makes.

Plan You can *draw a picture* to solve this problem.

Solve _____ sits at the desk at (5,2). _____ sits at the desk at (5,4). _____ sits at the desk at (1,4). Draw José's path on the grid. What shape is it? ________________________

Answer: In order, José walks to the desks of _________, ___________, and ___________. When he returns to his desk, the shape of his path is a __________________.

Look Back How can you solve this problem another way?

____

Practice Test Questions

Find Out • Plan • Solve • Look Back

Use the map for problems 1 and 2.

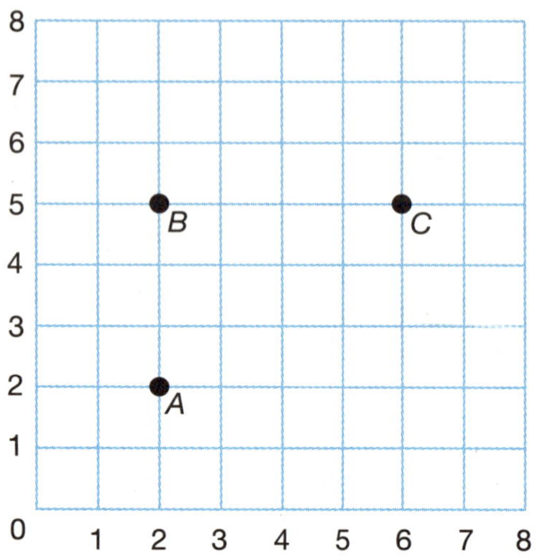

1. Which ordered pair names point *C*? (Hint: *Act out.*)

A (5,6)
B (6,5)
C (5,5)
D (6,6)

2. Derek draws lines from *A* to *B*, from *B* to *C*, and from *C* to *A*. What figure does he draw? (Hint: *Draw a picture.*)

A square
B circle
C rectangle
D triangle

3. How would you find the ordered pair (4,6) on a map? Which of the sentences tells the correct way?

A Go 6 units to the right; go 4 units up.
B Go 6 units to the left; go 4 units up.
C Go 4 units to the right; go 6 units up.
D Go 4 units up; go 6 units down.

4. To find the ordered pair (10,12) on a map, how many units would you move to the right?

Record your answer. Then fill in the bubbles. Be sure to use the correct place value.

⓪	⓪
①	①
②	②
③	③
④	④
⑤	⑤
⑥	⑥
⑦	⑦
⑧	⑧
⑨	⑨

Use the seating plan for problems 5 and 6.

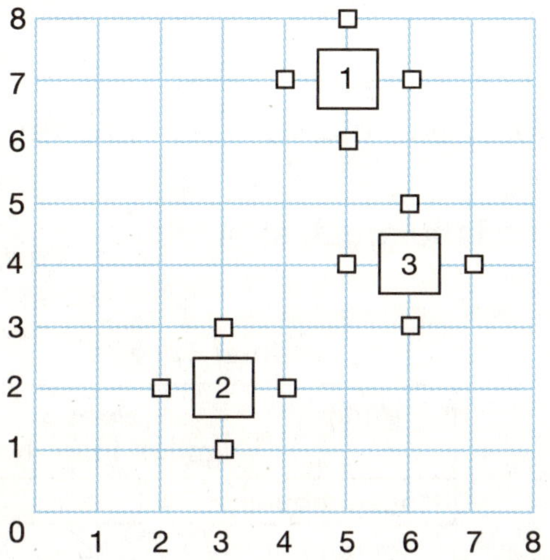

5. The map shows the seating plan for Mia's birthday lunch. Mia is sitting at the seat that is located at (4,7). At which table is Mia sitting? (Hint: *Solve a simpler problem.*)

6. Mia's cousin Serena is sitting at table 3. Give the ordered pairs of all the possible seats where Serena might be sitting. (Hint: *Act out.*)

7. Suppose you are standing on a map at point (3,1). You walk 5 blocks to your right. Then you walk 3 blocks up. Where are you standing now?

8. The map shows where Lisa put her stuffed animals on her bed.

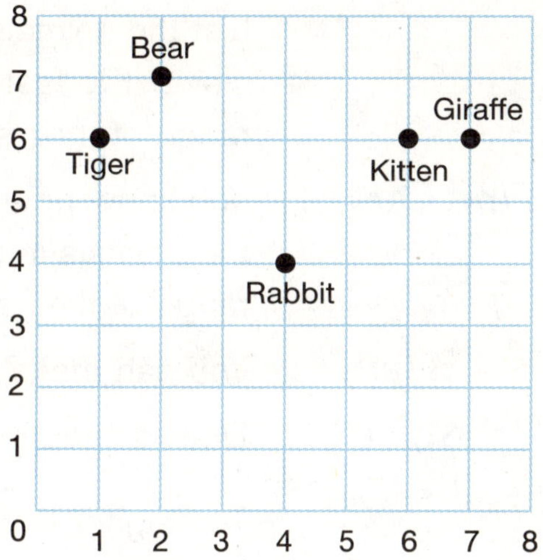

What stuffed animal is at (6,6)? Explain how you know.

____

____

____

____

Strategies: Picture • Model/Act Out • Organized List • Table/Graph • Pattern • Predict & Test Logical Thinking • Work Backward • Simpler Problem • Number Sentence

24 Analyze Data

You can use the mean, mode, and range to describe sets of numbers.

mean – the sum of all the numbers divided by the total number of numbers

mode – the number that occurs most often

range – the difference between the least and the greatest number

Example 1

The chart shows how fast some animals can run for a short distance.

List the animals in order from slowest to fastest. What is the range of their speeds?

Animal Speeds

Animal	Speed (miles in an hour)
House mouse	8
Cheetah	70
Lion	50
Chicken	9

Find Out

You know how fast each animal can run. You need to list the animals in order from slowest to fastest and find the range of their speeds.

Plan

You can *make a graph* to solve this problem.

Solve

Draw a graph of the speeds.

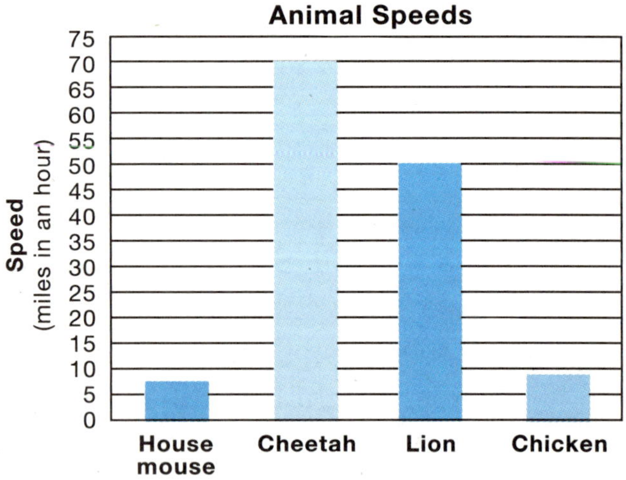

The higher the bar, the faster the animal runs. Subtract the slowest speed from the fastest speed to find the range.

$70 - 8 = 62$

Answer: The animals from slowest to fastest are house mouse, chicken, lion, and cheetah. The range is 62 miles in an hour.

Look Back

You can *make a table* to check your answer. Write the numbers in a place-value table to compare them. ✔

Try It

Example 2

Children in the third grade voted for their favorite ride at the town carnival. The pictograph below shows the results.

Favorite Carnival Rides	
Ride	**Votes**
Roller Coaster	✓✓✓✓✓
Carousel	✓✓✓
Bumper Cars	✓✓✓✓
Ferris Wheel	✓✓

Key: Each ✓ = 3 votes

What is the mean of the votes?

Find Out You know:
- The number of votes for each ride.
- The range is the difference between the greatest and least number of votes.

You need to find the mean.

Plan You can *write a number sentence* to solve the problem.

Solve First, find the number of votes for each ride.

Roller coaster: __________ votes

Carousel: __________ votes

Bumper cars: __________ votes

Ferris Wheel: __________ votes

Add to find the total. __________ votes

Write a number sentence you can use to find the mean.

__________ ÷ __________ = □

Solve the number sentence: □ = __________.

Answer: The mean of the votes is __________.

Look Back How can you solve this problem another way?

____

Practice Test Questions

Find Out • Plan • Solve • Look Back

1. The library just got 50 history books, 35 biographies, and 15 science books. Which graph shows the new books? (Hint: *Make a graph.*)

A **New Library Books**

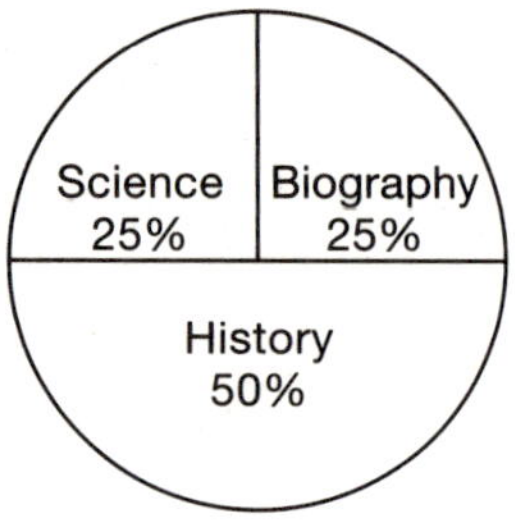

B **New Library Books**

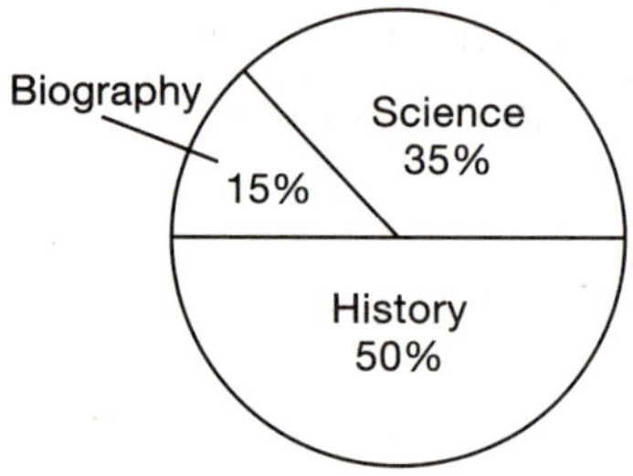

C **New Library Books**

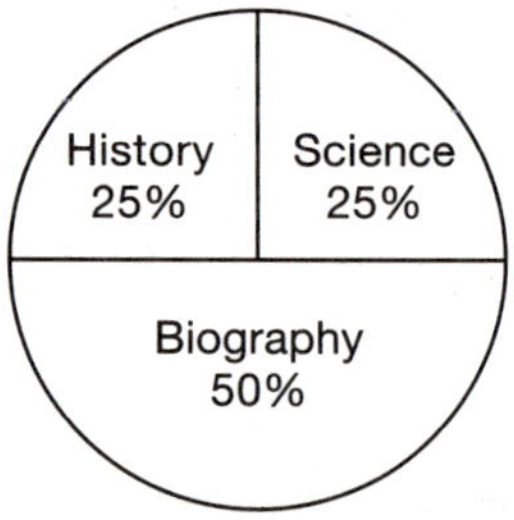

D **New Library Books**

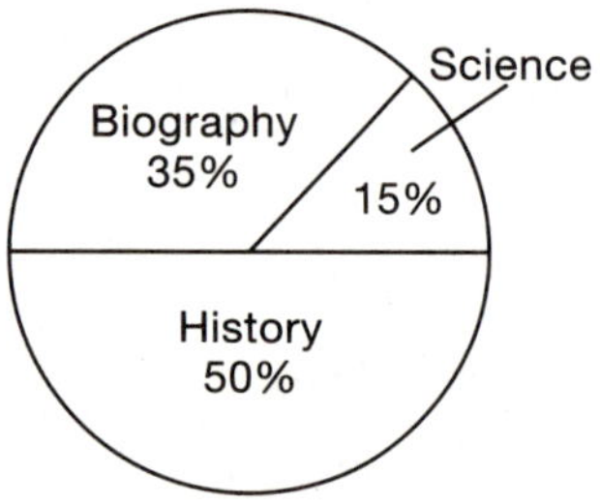

2. The ice-skating judges gave Harriet the following scores: 6, 3, 6, 4, 6. What is her mean score? (Hint: *Write a number sentence.*)

A 3

B 4

C 5

D 6

3. The ages of the Johnson children are 10, 17, 12, and 6. What is the range of their ages?

Record your answer. Then fill in the bubbles. Be sure to use the correct place value.

⓪	⓪
①	①
②	②
③	③
④	④
⑤	⑤
⑥	⑥
⑦	⑦
⑧	⑧
⑨	⑨

4. Ellen made stacks of all the pennies she found around the house.

What is the average number of pennies per stack. (Hint: *Act out.*)

5.

Snacks Price List	
Cotton Candy	50¢
Popcorn	95¢
Taffy Apple	80¢
Ice Cream	75¢

Jimmy has $1.75. Does he have enough to buy popcorn and ice cream? Explain. (Hint: *Predict and test.*)

____

____

____

____

6. Marie checked the price of a bike at several stores.

$68, $83, $76, $72, $70, $74

What is the median price?

Use the graph for problems 7 and 8.

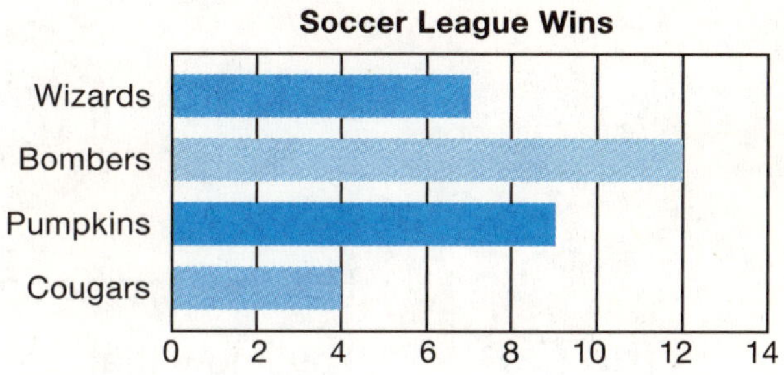

7. List the teams in order from the most wins to the fewest wins.

8. What is the average number of wins by all the teams in the soccer league?

Strategies: Picture • Model/Act Out • Organized List • Table/Graph • Pattern • Predict & Test • Logical Thinking • Work Backward • Simpler Problem • Number Sentence

25 Probability

Probability tells how likely it is that an event will happen. An event can be ***likely,*** but not certain. An event can be ***unlikely,*** but not impossible. Some events are ***equally likely.***

Example 1

Bryan dropped these two coins on the floor.

How many different ways could the coins land?

Find Out You know that Bryan dropped a dime and a nickel. You know that each coin can land on heads or tails. You want to find out all of the different ways the 2 coins can land.

Plan You can *make an organized list* to solve this problem.

Think: *I can list every possible way the coins can land.*

Solve

Dime	Nickel
Heads	Heads
Heads	Tails
Tails	Tails
Tails	Heads

Make a list of all the possible ways.

Count the number of different ways.

Answer: There are 4 possible ways for the coins to land.

Look Back You can make a model or act out to check your answer.

Use a dime and a nickel. Act out all the different ways and then count them. The answer will be 4 ways. ✔

Try It

Example 2

Connie and Darnell are playing a board game. They each roll a number cube numbered 1–6. Then they move their markers that number of spaces. They are trying to get to the END OF GAME space.

Connie is on the NOW YOU ARE CLOSE space. Is it more likely, less likely, or equally likely that Connie will win the game on her next roll?

Find Out

You know:
- the space Connie is on
- the numbers Connie can roll to win

You need to find out how likely it is that she will win on her next roll.

Plan

You can *draw a picture* to solve the problem.

Solve

Draw a picture to show the numbers that can be rolled using the number cube.

Circle the numbers that Connie can roll and still win the game.

Decide if rolling these numbers is more likely, less likely, or equally likely.

Answer: It is ____________________ that Connie will win the game on the next roll.

Look Back

How can you solve this problem another way?

____

Practice Test Questions

Find Out • Plan • Solve • Look Back

1. Marianne has a stack of 4 cards and a quarter.

She picks a card and tosses the quarter. How many possible outcomes are there? (Hint: *Make an organized list.*)

A 16　　**C** 4

B 8　　**D** 2

2. Which spinner is least likely to land on an even number? (Hint: *Use logical thinking.*)

A

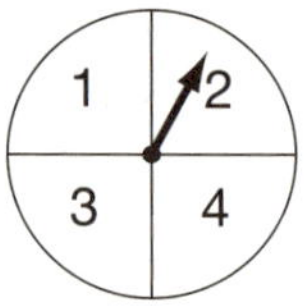

B

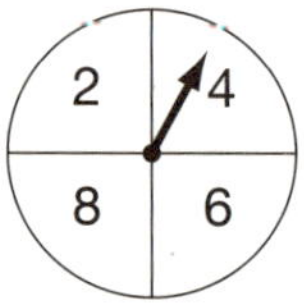

C

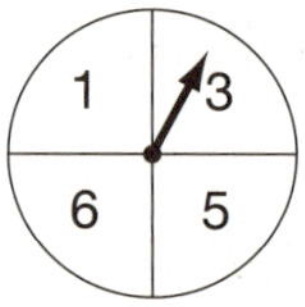

D

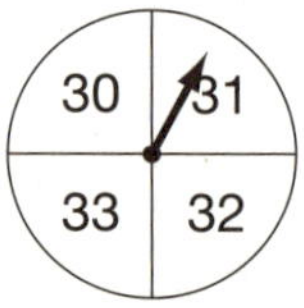

3. Ernesto has a stack of number cards.

2, 4, 4, 4, 2, 4, 4, 2, 4, 4

Which sentence is **true** about the next card Ernesto will pick?

A It is more likely he will pick a 2 than a 4.

B It is less likely he will pick a 4 than a 2.

C It is equally likely he will pick a 2 or a 4.

D It is more likely he will pick a 4 than a 2.

4. Harry tosses a coin and rolls a 1–6 number cube. How many possible outcomes are there?

Record your answer. Then fill in the bubbles. Be sure to use the correct place value.

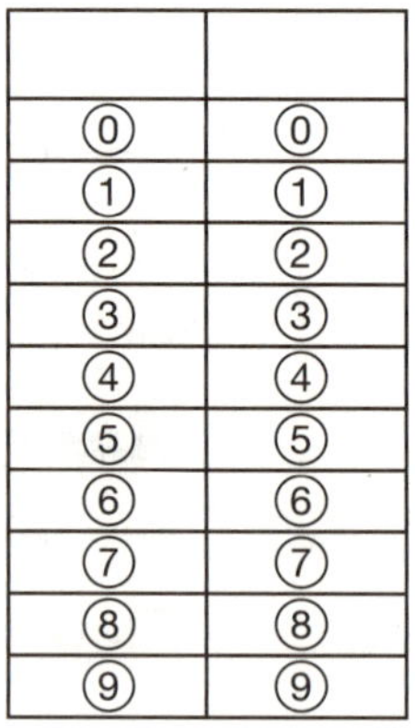

5. There are 3 red marbles, 4 blue marbles, and 5 white marbles in a bag. You pick a marble without looking. Is it more likely, less likely, or equally likely that you would pick a red or blue marble than a white marble? Explain why. (Hint: *Draw a picture.*)

6. Ariana wrote the letters of her name on slips of paper and put them in a bag. Which letter is she most likely to pick from the bag? Why? (Hint: *Draw a picture.*)

____

____

____

____

7. Justin has a large box where he keeps all his pattern blocks.

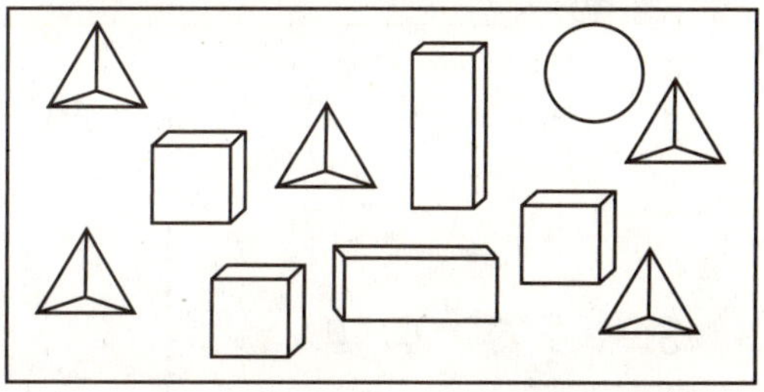

He picks a block from the box. Which block is he least likely to pick? (Hint: *Make a model or act out.*)

8. This is the spinner for the Spin-To-Win game.

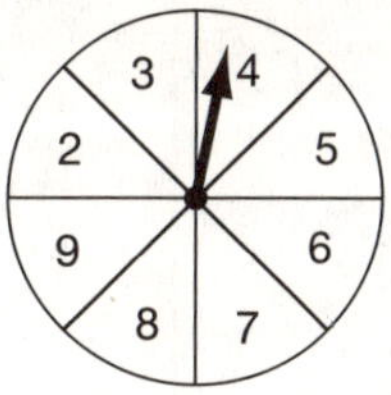

To win the game, you must land on 6, 7, 8, or 9. What are all the possible outcomes? Are you more likely, less likely, or equally likely to win the game? Explain your answer.

____

____

____

____

Strategies: Picture • Model/Act Out • Organized List • Table/Graph • Pattern • Predict & Test Logical Thinking • Work Backward • Simpler Problem • Number Sentence

Name: ______________________________

1. Ⓐ Ⓑ Ⓒ Ⓓ
2. Ⓐ Ⓑ Ⓒ Ⓓ
3. Ⓐ Ⓑ Ⓒ Ⓓ
4. Ⓐ Ⓑ Ⓒ Ⓓ
5. Ⓐ Ⓑ Ⓒ Ⓓ
6. Ⓐ Ⓑ Ⓒ Ⓓ
7. Ⓐ Ⓑ Ⓒ Ⓓ
8. Ⓐ Ⓑ Ⓒ Ⓓ
9. Ⓐ Ⓑ Ⓒ Ⓓ
10. Ⓐ Ⓑ Ⓒ Ⓓ
11. Ⓐ Ⓑ Ⓒ Ⓓ
12. Ⓐ Ⓑ Ⓒ Ⓓ
13. Ⓐ Ⓑ Ⓒ Ⓓ
14. Ⓐ Ⓑ Ⓒ Ⓓ
15. Ⓐ Ⓑ Ⓒ Ⓓ
16. Ⓐ Ⓑ Ⓒ Ⓓ
17. Ⓐ Ⓑ Ⓒ Ⓓ
18. Ⓐ Ⓑ Ⓒ Ⓓ
19. Ⓐ Ⓑ Ⓒ Ⓓ

20.

⓪	⓪
①	①
②	②
③	③
④	④
⑤	⑤
⑥	⑥
⑦	⑦
⑧	⑧
⑨	⑨

21. ____________________

22. ____________________

23. ____________________

24. ____________________

25. ____________________

26. ____________________

27. ____________________

28.

⓪	⓪
①	①
②	②
③	③
④	④
⑤	⑤
⑥	⑥
⑦	⑦
⑧	⑧
⑨	⑨

29. ____________________

30. ____________________

31.

⓪	⓪
①	①
②	②
③	③
④	④
⑤	⑤
⑥	⑥
⑦	⑦
⑧	⑧
⑨	⑨

32. ____________________

33. ____________________

34. ____________________

35. ____________________

Test-Taking Tips

When you take a test, read each question carefully. Answer all the questions. Always ask yourself if your answer makes sense. If you have time, check your answers before you hand in the test.

Tips for Multiple-Choice Questions

1. Read all the choices before you choose an answer.
2. If you're not sure of the right answer, cross out wrong answers. Then make the best guess you can.
3. Mark the correct answer. Use the answer sheet or circle the letter of the correct answer on the test page.
4. If you change an answer, erase the answer you don't want.

Tips for Using the Bubble Grid or Writing an Answer

1. After you decide on the answer, check it. Usually, you can start with your answer and work backward.
2. For the bubble grid, write the answer in the top row. Then fill in the bubbles that match. Be sure to use the correct place value.
3. When you write out an answer, write clearly. If you change the answer, neatly erase or cross out the answer you don't want.

How to Record your Answers

Sample question: 1. There are 12 inches in 1 foot.
How many inches are there in 3 feet?

Working on the test sheet

A 12 inches

B 1 yard

(C) 36 inches

D 12 feet

Working on the answer sheet

Ⓐ Ⓑ ● Ⓓ

Fill in the circle completely.

Writing out the answer

36 inches

$12 \times 3 = 36$

Sometimes you have to show your work.

Using the bubble grid

3	6
⓪	⓪
①	①
②	②
●	③
④	④
⑤	⑤
⑥	●
⑦	⑦
⑧	⑧
⑨	⑨

Practice Test

Find Out • Plan • Solve • Look Back

1.

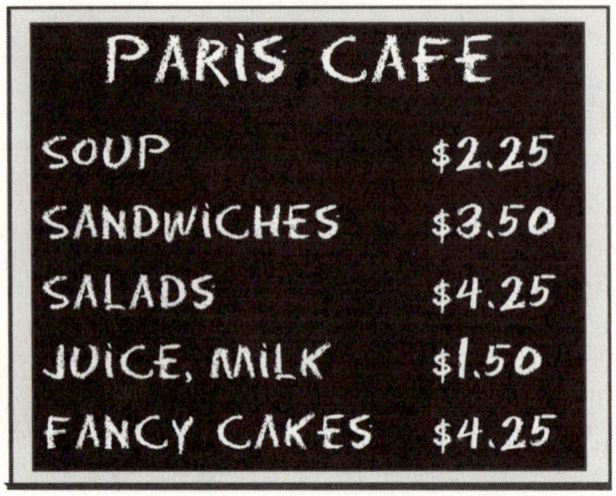

Mrs. Harris took her family to the Paris Café. There were 4 adults and 8 children. Each person had a sandwich and juice or milk. What was the total cost?

A $37.00

B $40.00

C $42.00

D $60.00

2. Linda and her sister Susan have 18 beanbag animals. Susan has 4 more than Linda. How many beanbag animals does Linda have?

A 18 animals

B 14 animals

C 11 animals

D 7 animals

3. Alden is saving to buy a model rocket. The graph shows the total of his savings each week.

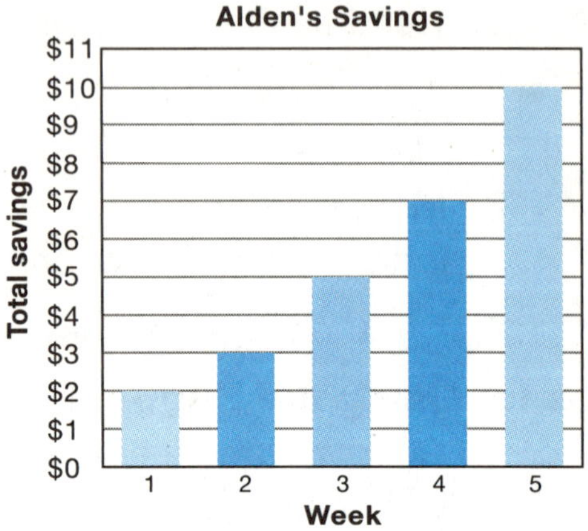

Which sentence is **true**?

A Alden saves $1 each week.

B Alden saves $2 each week.

C Alden saves $3 each week.

D Alden saves $4 each week.

4. A chameleon is a lizard that changes color. Its tongue is 2 times the length of its body. Which number sentence would you use to find the length of the tongue of a 3-inch chameleon?

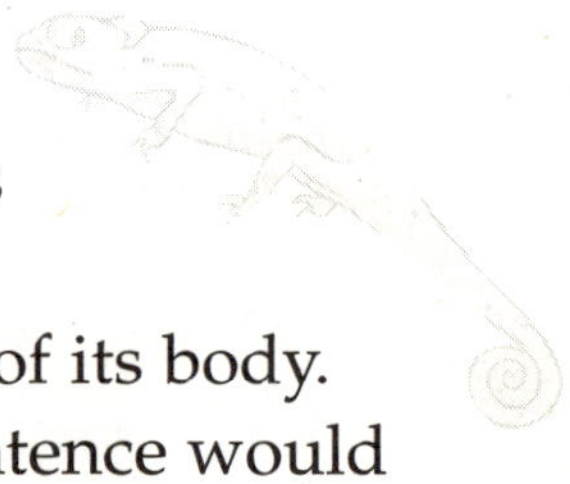

A $2 \times 3 = \square$

B $3 \div 2 = \square$

C $\square + 2 = 3$

D $3 + 2 = \square$

5. Raul bought a package of cheese. He ate $\frac{1}{4}$ pound at lunch. How much cheese is left?

What else do you need to solve this problem?

A the number of sandwiches made

B the number of Raul's brothers

C how much bread Raul used

D how much cheese is in a full package

6. Every morning, the first bus stops at the corner at 6:00 A.M. A bus stops there every 30 minutes. At what time will the fifth bus stop at the corner?

60 minutes = 1 hour

A 7:30 A.M.

B 8:00 A.M.

C 8:30 A.M.

D 9:00 A.M.

7. Hilary ran $1\frac{1}{4}$ miles. Joel ran $1\frac{2}{8}$ miles. Alma ran $1\frac{1}{4}$ miles. Paul ran $1\frac{5}{8}$ miles. Who ran the longest distance?

A Alma

B Hilary

C Joel

D Paul

8. The circle graph shows how Kate spends her weekly allowance.

Kate's Budget

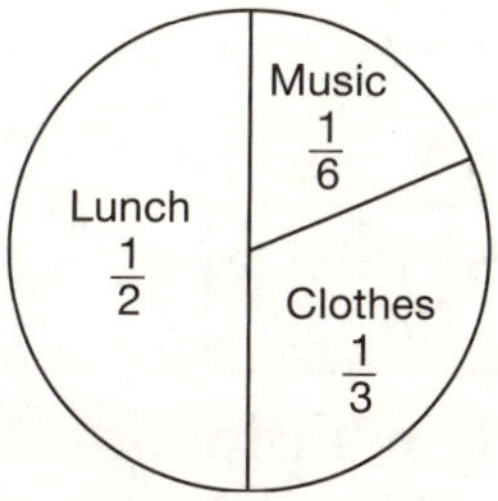

Which sentence is **true**?

A Kate spends more on clothes than on music.

B Kate spends more on clothes than on lunch.

C Kate spends more on music than on lunch.

D Kate spends the same amount on music and clothes.

9. The last movie of the day ended at 10:30 P.M. The movie ran for 1 hour 45 minutes. Which clock shows the time the movie started?

A

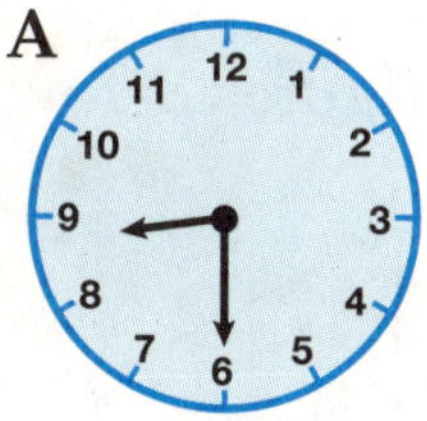

C

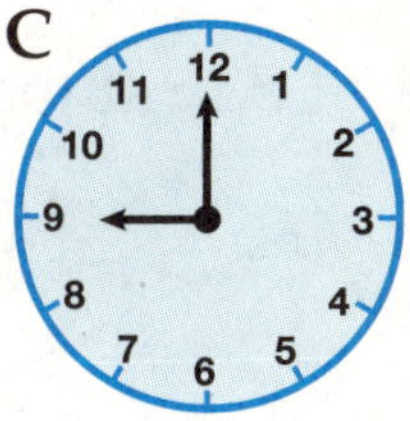

B

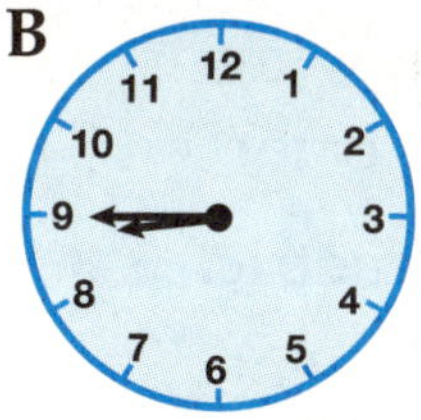

D

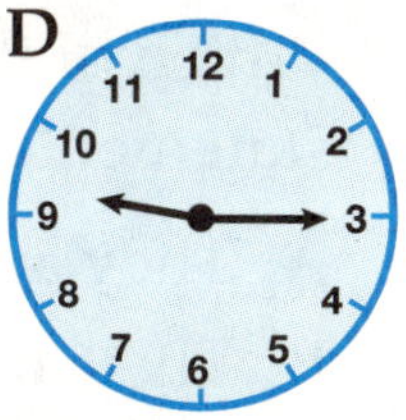

10. Charles drinks $1\frac{1}{4}$ cups of milk at breakfast, $2\frac{1}{4}$ cups at lunch, and $1\frac{3}{4}$ cups at dinner. How much milk does Charles drink in all?

2 cups = 1 pint
4 cups = 1 quart

A less than a pint
B less than a quart
C more than a quart
D more than 2 quarts

11. Kristi is mailing a gift to her aunt. The gift weighs 6.5 pounds. The mailing box weighs 0.7 pound. How much will the package weigh in all?

A 5.8 pounds
B 6.2 pounds
C 7.2 pounds
D 8.2 pounds

12. Look at the two figures below.

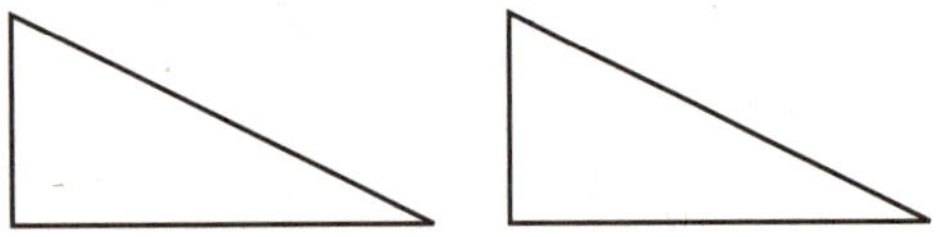

Which of the following statements is **not** true?

A They are symmetric.
B They are congruent.
C They are triangles.
D Each has 3 angles and 3 sides.

13. Which of the following number sentences is **true**?

A $67 - 59 < 40 - 35$
B $53 - 21 = 29 + 12$
C $36 + 55 > 57 + 28$
D $42 + 17 > 83 + 51$

14. Mark walks the length of one side of a square pool. He counts 20 steps. How many steps will Mark take to walk around the pool?

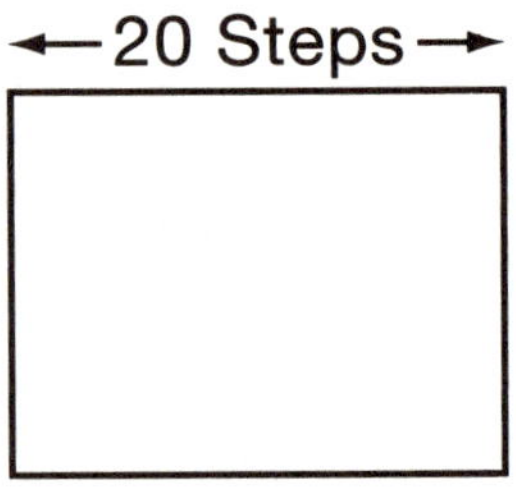

A 40 steps
B 80 steps
C 100 steps
D 400 steps

15. Theo bought presents for 5 friends. He paid $4 for each. He also paid $12 to have all of them packed and mailed. Which of the following shows how to find Theo's total cost?

A $5 + 4 + 12 = 21$
B $5 - 4 + 12 = 13$
C $5 \times 4 = 20;\ 20 - 12 = 8$
D $5 \times 4 = 20;\ 20 + 12 = 32$

16. The map below shows where four students live.

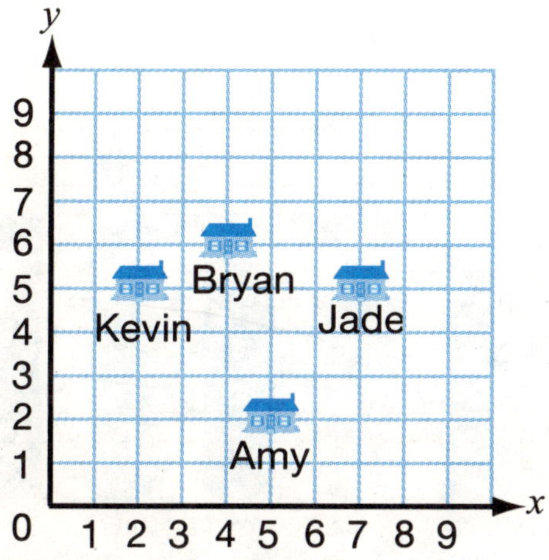

Who lives at (5,2)?

A Amy

B Bryan

C Jade

D Kevin

17. Elliott buys a pair of sneakers for $27 and a pair of jeans for $11. His mother gives the clerk a $50 bill. How much change will she get?

A $12

B $23

C $38

D $39

18. Sonia dips the end of a prism in paint and presses it on paper. This is the shape she gets.

Which prism did she use?

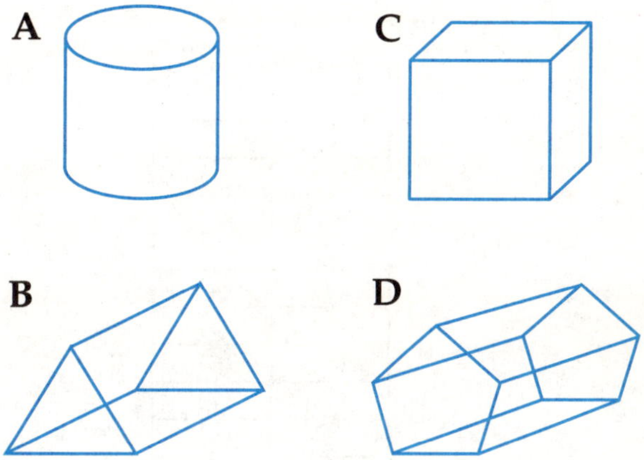

19. Evan spins the spinner. Which statement about the spinner is **true**?

A It is more likely to land on a number greater than 6 than on a number less than 6.

B It is more likely to land on 3 than on 9.

C It is more likely to land on an even number than on an odd number.

D It is more likely to land on an odd number than on an even number.

20. Thomas Jefferson was born in 1743. Abraham Lincoln was born in 1809. How old was Thomas Jefferson when Abraham Lincoln was born?

Record your answer. Then fill in the bubbles. Be sure to use the correct place value.

⓪	⓪
①	①
②	②
③	③
④	④
⑤	⑤
⑥	⑥
⑦	⑦
⑧	⑧
⑨	⑨

21. Last week, Morgan swam 8 laps every day. At the end of the week, she was still 9 laps short of her goal. What was her goal? Explain how you got your answer.

____

____

____

22. A woodpecker can peck 20 times each second. How many times can the woodpecker peck in 5 seconds?

23. The bar graph shows the number of basketballs sold at the Sport House in one week.

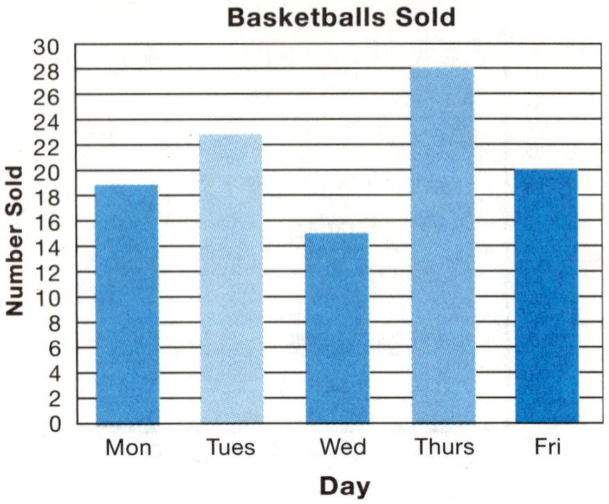

How many more basketballs were sold on Thursday than on Wednesday?

24. The tusks of an African elephant grow about 7 inches per year. About how long would it take for an elephant to grow tusks 35 inches long?

25. Toby bought these items.

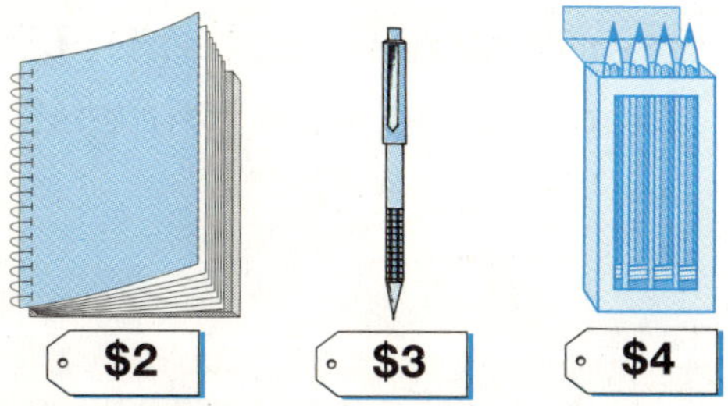

She got $11 change from the clerk. How much did she give the clerk? Explain how you got your answer.

26. Parker jogged 5.3 kilometers on Monday and 6.2 kilometers on Wednesday. How much farther did he jog on Wednesday?

27. An adult Asian elephant can weigh 8,250 pounds. At birth, it can weigh 189 pounds. How much weight can an Asian elephant gain in its lifetime?

28. There are 6 small boxes of raisins in a package. There are 4 small bags of nuts in a box. Billy wants the same number of raisin boxes as bags of nuts. What is the fewest number of each he can buy?

Record your answer. Then fill in the bubbles. Be sure to use the correct place value.

⓪	⓪
①	①
②	②
③	③
④	④
⑤	⑤
⑥	⑥
⑦	⑦
⑧	⑧
⑨	⑨

29. Jay writes these numbers on cards.

1	2	3	4	5

He puts the cards in a bag and draws one without looking. Is he more likely to pick an odd or an even number? Explain.

30. A school buys 56 computers. They will be shared equally among 4 classrooms. How many computers will each class get?

31. At 8 A.M. the temperature was 48 degrees. It went up 6 degrees by noon. Then it fell 9 degrees by 8 P.M. How many degrees was the temperature at 8 P.M.?

Record your answer. Then fill in the bubbles. Be sure to use the correct place value.

⓪	⓪
①	①
②	②
③	③
④	④
⑤	⑤
⑥	⑥
⑦	⑦
⑧	⑧
⑨	⑨

32. Scott puts his coins in a piggy bank every night. Today, he has 9 quarters, 23 dimes, 14 nickels, and 17 pennies in the bank. How much money does Scott have?

33. A new pair of ice skates costs $32 more than a used pair. A used pair of skates costs $15. How much does a new pair cost?

34. Which 2 figures are congruent?

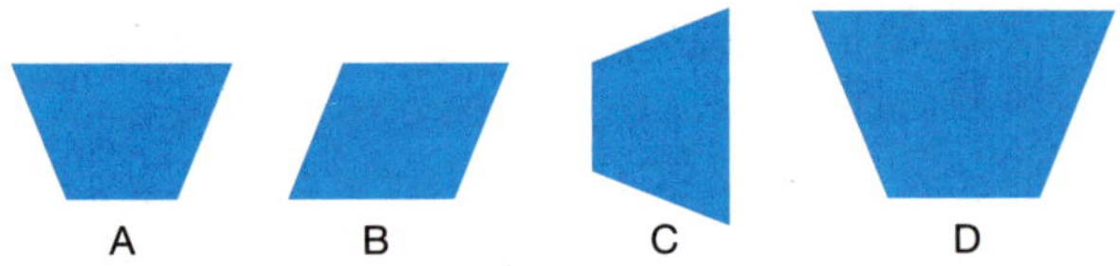

35. Maria is playing a board game. On each turn, she moves ahead 8 spaces and back 2 spaces. How many turns will it take Maria to move ahead a total of 18 spaces? Show how you got your answer.